Some Notes On Shipbuilding and Shipping In Colonial Virginia

By

CERINDA W. EVANS
Librarian Emeritus, The Mariners Museum
Newport News, Virginia

Virginia 350th Anniversary Celebration Corporation
Williamsburg, Virginia
1957

Jamestown 350th Anniversary
Historical Booklet, Number 22

AS CONCERNING SHIPS

It is that which everyone knoweth and can say

> They are our Weapons
> They are our Armaments
> They are our Strength
> They are our Pleasures
> They are our Defence
> They are our Profit
> The Subject by them is made rich
> The Kingdom through them, strong
> The Prince in them is mighty
> In a word: By them in a manner we live
> The Kingdom is, the King reigneth.
> (From *The Trades Increase*, London, 1615)

SHIPBUILDING AND SHIPPING

The Dugout Canoe

Various types of watercraft used in Colonial Virginia have been mentioned in the records. The dugout canoe of the Indians was found by the settlers upon arrival, and was one of the chief means of transportation until the colony was firmly established. It is of great importance in the history of transportation from its use in pre-history to its use in the world today. From the dugout have come the piragua, Rose's tobacco boat, and the Chesapeake Bay canoe and bugeye as we see them today.

The first boats in use by the colony in addition to the Indian canoe were ships' boats—barges, long-boats, and others. A shallop brought over in sections was fitted together and used in the first explorations. As the years went by, however, "almost every planter, great and small, had a boat of one kind or another. Canoes, bateaux, punts, piraguas, shallops, flats, pinnaces, sloops, appear with monotonous regularity in the seventeenth and eighteenth century records of Virginia and Maryland."

Little is known about the construction of boats in the colony except the log canoe. A long and thick tree was chosen according to the size of the boat desired, and a fire made on the ground around its base. The fire was kept burning until the tree had fallen. Then burning off the top and boughs, the trunk was raised upon poles laid over crosswise on forked posts so as to work at a comfortable height. The bark was removed with shells; gum and rosin spread on the upper side to the length desired and set on fire. By alternately burning and scraping, the log was hollowed out to the desired depth and width. The ends were scraped off and rounded for smooth navigating.

Captain John Smith, who had a number of occasions to use the canoe, wrote that some were an elne deep (forty-five inches), and forty or fifty feet in length; some would bear forty men, but the most ordinary were smaller and carried ten, twenty, or thirty men. "Instead of oars, they use paddles or sticks with which they will row faster than our barges." Additional space and graceful lines in the canoes were secured by spreading the sides. To do this, the hollowed log was filled with water and heated by dropping in hot stones until the wood became soft enough to bend into the desired shape by forcing the sides apart with sticks of different lengths and allowed to harden.

The tools with which the Indians built their boats and used for other purposes, were tomahawks of stone sharpened at one end or both, or one end was rounded off for use as a hammer. A circular indentation was made in the center to secure the tomahawk to the handle. Another method of fitting the stone tomahawk to a handle was to cut off the head of a young tree, and as if to graft it, a notch was made into which the head of the hatchet was inserted. After some time, the tree by growing together kept the hatchet so fixed that it could not come out. Then the tree was cut to such a length as to make a good handle. Another method in use was that of binding the stones to the ends of sticks and gluing them there with rosin.

Some colonists did not hesitate to take the canoes from the Indians, which they may or may not have returned. On one occasion the King of Rappahanna demanded the return of a canoe, which was restored. Among the first laws of the General Assembly was that for the protection of the Indians, enacted in August, 1619: "He that shall take away by violence or stealth any canoe or other things from the Indians, shall make valuable restitution to the said Indians, and shall forfeit, if he be a freeholder, five pounds; if a servant, forty shillings or endure a whipping."

A story of an Indian and his canoe was told by John Pory, Secretary of Virginia, after he had visited the Eastern Shore. "Wamanato, a friendly Indian, presented me with twelve bever skins and a canow which I requited with such things to his con-

2

tent, that he promised to keep them whilst he lived, and berie them with him being dead."

Several writers of boatbuilding have expressed the thought that the evolution of the Chesapeake Bay canoe and the Chesapeake Bay bugeye from the Indian dugout canoe was one of the most interesting developments in the history of shipbuilding. M. V. Brewington, in his *Chesapeake Bay: A Pictorial Maritime History*, says of this development: "The white man's superior knowledge of small craft soon indicated changes which would improve the canoe: sharp ends would make her easier to propel and more seaworthy; broader beam and a keel would increase stability; sail would lessen the work of getting from place to place. Sharpening the bow and stern was a simple matter; the increased beam was difficult because no single tree could provide the needed width. In time, the settler learned to join two or more trees together to give the beam desired. He learned how to add topsides, first of hewn logs, later of sawed plank. A keel was added and a sailing rig. After the centerboard was invented, it took the place of a keel. . . ."

"But the culmination of the simple, single log, trough-shaped Indian dugout was the bugeye, a complex vessel as much as eighty-five feet in length. There was an intermediate step between the canoe and the bugeye, the brogan, a large canoe, partially decked, with a cuddy forward in which a couple of men could sleep and cook. . . . The earliest known use of the name "bugeye" was in 1868, but doubtless the word was not coined upon the first appearance of the vessel itself. . . . In essence the bugeye was a large canoe, fully decked, with a fixed rig following that of the brogan. There were full accommodations for the crew which, because the vessel was built for oyster dredging, needed to be comparatively large. . . . Throughout the course of development from canoe to bugeye, the original dugout log bottom was always apparent in this most truly American craft."

The smallest of the three vessels that reached Virginia in April, 1607, was the little pinnace *Discovery,* a favorite type of small vessel in that period. The first English vessel known to have been built in the New World was a pinnace. A colonizing expedition to Raleigh's colony on Roanoke Island left Plymouth, England, on April 9, 1585, with a fleet of five vessels and two pinnaces attached as tenders. A storm sank the tender to the *Tiger,* Sir Richard Grenville's flagship. On the 15th of May, the fleet came to anchor in the Bay of Mosquetal (Mosquito), and a landing was made at St. John on the Island of Puerto Rico. Here an encampment was made to give the men time to refresh themselves and to build a new pinnace for the *Tiger.* A forge was set up to make the nails, and trees were cut and hauled to camp on a low four-wheeled truck for the boat's timber. The ship's carpenters made speedy headway, launching and rigging the pinnace in ten days. They set sail from St. John on the 29th of May, the new pinnace carrying twenty men and, on the 27th of July, anchored at Hatoraske on the way to Roanoke.

The second English vessel known to have been built in North America was also a pinnace. The members of the second colony of Virginia left Plymouth, England, on the last day of May, 1607, under command of Captain George Popham, and located at "Sagadahoc in Virginia" at the mouth of the Kennebec River. There they set up fortifications which they called Fort St. George. After finishing the fort, "the carpenter framed a pretty pinnace of about thirty tons which they called *Virginia,* the shipwright being one Digby of London." This little vessel is known to have made two voyages across the Atlantic.

On June 7, 1609, a fleet of seven ships and two pinnaces left Plymouth, England, for Jamestown. After a few days out, one of the pinnaces returned to England, but the other, the little *Virginia,* remained with the fleet as the tender to the flagship

4

Sea Venture. Sir Thomas Gates, Lieutenant Governor under Lord De La Warr, and Sir George Somers, Admiral of the fleet, embarked on the *Sea Venture,* commanded by Captain Christopher Newport, Vice Admiral. These three men were leaders of the expedition and in order to avoid any dispute as to precedence, they agreed—very unwisely, it was disclosed—to sail on the same ship "with several commissions sealed, successively to take place one after another, considering the uncertainty of human life."

Wreck of the *Sea Venture*

On July 28, a violent storm arose which separated the *Sea Venture* from the rest of the fleet. This "dreadful tempest" was the tail of a West Indies hurricane and lasted four days and nights. An account of it written in 1610, by William Strachey, secretary to Lord De La Warr, and a passenger on the ship, is said to be one of the finest descriptions of a storm in all literature, and led to the writing of *The Tempest* by Shakespeare. The letter was written to a person unknown, addressed as "Excellent Lady." Some excerpts are given herewith.

When on S. James his day, July 24, being Monday . . . the clouds gathering thicke upon us and the wind singing and whistling most unusually, which made us to cast off our pinnace towing the same until then asterne, a dreadful storm and hideous, began to blow from out the north-east, which swelling, and roaring, as it were by fitts, some hours with more violence than others, at length beat all light from heaven, which like a hell of darkness turned black upon us, so much the more fuller of horror, as in such cases horror and fear use to overrunne the troubled, and overmastered sences of all, which, taken up with amazement, the eares lay so sensible to the terrible cries, and murmurs of the winds, and distractions of our company . . . For foure and twenty houres the storme in a restless tumult, had blown so exceedingly, as we could not apprehend in our imaginations any possibility of greater violence, yet did wee still find it, not only more terrible, but more constant, fury added to fury, and one storm urging a second more outrageous than the former; whether it so wrought upon our feares . . . as made us look one upon the other with troubled hearts and panting bosoms; our clamours drowned in the windes, and the windes in thunder. Prayers might well be in the heart and lips, but drowned in the outcries of the officers, nothing heard that could give comfort, nothing seen that might encourage hope . . . The sea swelled above the clouds, and gave battell unto Heaven. It could not be said to raine, the waters like whole rivers did flood in the ayre . . . The winds spake more loud and grew more tumultuous and malignant. What shall I say? Winds and seas were as mad as fury and rage could make them . . . There was not a moment in which the sudden splitting or instant oversetting of the ship was not expected. Howbeit this was not all; it pleased God to bring a greater affliction yet upon us; for in the beginning of the storm, we had received likewise a mighty leake. And the ship in every joint almost, having spued out her okam, before we were aware . . . was growne five foote suddenly deep with water above her ballast, and we almost drowned within, whilst we sat looking when to perish from above. This imparting no less terror than danger ran through the whole ship with much fright and amazement, startled and turned the blood and took down the braves of the most hardy mariner of them all . . . The leake which drunk in our greatest seas, and took in our destruction fastest could not then be found nor ever was by any labour, counsell or search . . . Every man came duely upon his watch . . . working with tyred bodies and wasted spirits three days and foure nights destitute of outward comfort, and desperate of any deliverance . . . During all this time the Heavens looked so black upon us that it was not possible the elevation of the pole might be observed; nor a starre by night, not a sun beame by day was to be seene. Onely upon Thursday night, Sir George Somers being upon the watch, had an apparition of a little round light like a faint starre, trembling and streaming along with a sparkeling blaze, halfe the height upon the main mast, and shooting sometimes from shroud to shroud, tempting to settle as it were, upon any of the foure shroudes . . . half the night it kept with us; running sometimes along the main yard to the very end, and then returning. At which, Sir George Somers called divers about him, and showed them the same . . . It did not light us any whit the more to our known way, who ran now as hoodwinked men, at all adventures, sometimes north and north-east, then north and by west, and in an instant varying two or three points, and sometimes half the compass . . . It being now Friday, the fourth morning, it wanted little, but that there had been a general determination to have shut up hatches, and commending our sinfull soules to God, committed the ship to the mercy of the sea. Surely, that night we must have done it, and that night had we then perished: but see the goodnesse and sweet introduction of better hope, by our merciful God given unto us. Sir George Somers, when no man dreamed of such happiness, had discovered and cried land!

The storm drove the ship toward the dangerous and dreaded islands of Bermuda. Nearing the shore, the ship was caught between rocks as in a vise and held there while all the one hundred and fifty persons reached the shore in safety. As soon as they were conveniently settled, after the landing, the long boat was fitted up in the fashion of a pinnace with a little deck made of the hatches of the wrecked ship, so close that no water could enter, and with a crew of six sailors, using sails and oars, Thomas Whittingham, the cape merchant, and Henry Ravens, the master's mate, as pilot, the boat sailed for Virginia. It was hoped, when news reached Jamestown of the safe landing of the passengers from the wrecked *Sea Venture* on Bermuda, that a ship or pinnace from the fleet in Virginia would be sent to take them home, but the long boat was never heard from again.

Building the *Deliverance* and the *Patience*

While waiting for help from Virginia, Sir George Somers and Sir Thomas Gates decided to build a pinnace, in case of need. The work was put in charge of Richard Frobisher, an experienced shipwright. The only wood on the island that could be used for timber was cedar and that was rather poor, being too brittle for making good planks. The pinnace's beams were all of oak from the wrecked ship, as were some planks in her bow, all the rest was of cedar. The keel was laid on the 28th of August, 1609, and on the 26th of February, calking had begun. Old cables that had been preserved furnished the oakum. One barrel of pitch and another of tar had been saved. Lime was made of wilk shells and a hard white stone, which were burned in a kiln, slaked with fresh water, and tempered with tortoise oil. She was forty feet long at the keel, nineteen feet broad at the beam, had a six-foot floor, her rake forward being fourteen feet, her rake aft from the top of her post (which was twelve feet long) was three feet; she was eight feet deep under her beam, four feet and a half between decks, with a rising of half a foot more under her fore-castle, the purpose being to scour the deck with small shot if an enemy should come aboard. She had a fall of eighteen inches aft to make her steerage and her great cabin larger; her steerage was five feet long and six feet high with a closed gallery right aft, having a window on each side, and two right aft. She was of some eighty tons burden.

On the 30th of March, the pinnace was launched, unrigged, and towed to "a little round island" nearer the ponds and wells of fresh water, with easier access to the sea, the channel there being deep enough to float her when masts, sails and all her trim had been placed on her. "When she began to swim (upon her launching) our Governor called her *The Deliverance*."

Late in November, and still with no word from Virginia, Sir George Somers became convinced that the pinnace which Frobisher was building would not be sufficient to transport all the men, women, and children from Bermuda to Virginia. He consulted with Sir Thomas Gates, the Governor, who approved his plan of building another pinnace. He would take two carpenters and twenty men with him to the main island where with instruction from Frobisher, "he would quickly frame up another little bark, for the better sitting and convenience of our people." The Governor granted him all the things he desired, all such tools and instruments, and twenty of the ablest and stoutest men of the company to hew planks and square timber. The keel laid was twenty-nine feet in length, the beam fifteen feet and a half; she was eight feet deep and drew six feet of water, and was of thirty tons capacity. Sir George Somers launched her on the last day of April, giving her the name of *Patience,* and brought her from the building bay in the main island, into the channel where the *Deliverance* was moored.

After nine months on the islands, these fearless and undaunted men, with a stout determination to finish the voyage they had begun nine months before, set sail in the two pinnaces on May 10, 1610, and after eleven days, arrived at Point Comfort. "On the three and twentieth day of May, we cast our anchor before Jamestown."

Boatbuilding Before 1612

The few available records of early boatbuilding in the Virginia colony differ so materially that one cannot make a statement as to number or kind of vessels with any degree of accuracy. That the first vessel constructed in Virginia was built earlier than the year 1611, and was of twelve or thirteen tons capacity, seems to be an accepted fact as given in the Spaniard Molina's *Report of a Voyage to Virginia in 1611.* The report also referred to a galley of twenty-five benches being built there.

In his *Short Relation* to the Council of the Virginia Company in June, 1611, Lord De La Warr spoke regretfully of the fact that the three forts he had erected near Point Comfort were not properly manned because of a lack of boats, there being but two, and one barge in all the colony. The fishing, too, had been hindered because of this shortage. No mention was made of the galley that was said to have been in the process of construction.

Argall's Shipyard at Point Comfort

In a letter to Nicholas Hawes, written in June, 1613, Samuel Argall (later Sir Samuel Argall) tells of a voyage to Virginia in 1612, and some of his activities there. On the 17th of September, he arrived at Point Comfort with sixty-two men on the ship *Treasurer,* his course being fifty leagues northward of the Azores. From the day of his arrival until the first of November, he spent the time in helping to repair such ships and boats as he found there "decayed for lack of pitch and tarre." About the first of November, he carried Sir Thomas Dale in the *Deliverance* to Sir Thomas Smith's Island to have his opinion about inhabiting it. They found an abundance of fish there, "very great cod" which they caught in water five fathom deep. They planned to get a great quantity in the summer of 1613, and hoped to find safe passage there for boats and barges by "a cut out of the bottom of our bay into De La Warr Bay." This is an early mention of the need for a canal connecting these two bays. That the Sir Thomas Smith's Island referred to was not the island known by that name lying near Cape Charles is evident from the reference to large cod fish caught there, and the desire for a passage between the bays for a shorter route.

Argall sailed from Point Comfort on the first of December and entered Pembroke, now Rappahannock, River where he met the king of Pastancie, who told him the Indians were his very great friends and had a good store of corn for him, as they had provided the year before. He carried his ship to the king's town and there built a stout shallop to take the corn aboard. After concluding a peace with other divers Indian lords, and giving and taking hostages, Argall hastened to Jamestown with 1100 bushels of corn, which he delivered to the storehouses there, besides the 300 bushels he retained for the use of his own company. As soon as he had unloaded the corn, Argall set his men to work felling timber and hewing boards with which to build a "frigat." He left this vessel half finished in the hands of his carpenters at Point Comfort in order to make another voyage to Pembroke River, and so discovered the head of it. Upon learning that Pocahontas was with the King of Patowomack, he devised a stratagem by which she was captured. Pocahontas was taken to Jamestown and delivered to the protection of Sir Thomas Gates, who hastened to conclude with Powhatan, her father, a peace based upon the terms demanded by Argall. Argall returned to Point Comfort and "went forward with his frigat and finished her." He sent a "ginge" of men with her to Cape Charles, to get fish and transport them to "Henries Town" (Henrico). Another gang was employed to fell timber and cleave planks to build a fishing boat. Argall himself, with a third gang, left in the shallop on the first day of May to explore the east side of the Bay. Having explored along the shore for some forty leagues northward, he returned on the 12th of May, fitted his ship and built a fishing boat, and made ready to take the first opportunity for a fishing voyage.

Other Voyages of Argall

Samuel Argall is said to have achieved lasting fame as one of England's maritime pioneers by establishing a shorter route to Virginia from England in 1609, although Batholomew Gosnold took that route in 1602, and Martin Pring did so in 1603. The usual course led by way of the Canaries to the Island of Puerto Rico in the West Indies, the route of Columbus, a long, circuitous pathway exposed to pirates and interference from Spain. Argall made the round trip by the shorter route in five months. However, the shorter route did not supplant entirely the longer southern route for several decades.

Argall accompanied Lord De La Warr to Virginia in 1610, to point out the northern route. While in Virginia, he was sent with Sir George Somers to Bermuda with two pinnaces to get a supply of hogs and other provisions for the colony. In a storm, Argall lost sight of Sir George's pinnace and failed to locate Bermuda; so he changed his course toward the north and went to Sagadahoc and Cape Cod where he procured a large cargo of fish, which he brought to Jamestown. Sir George Somers reached Bermuda, but died there on November 9, 1610. Argall was then sent by Lord De La Warr to the river Patawomeke to trade with the Indians for corn, where he rescued the English boy, Henry Spelman, who had been living with the Indians. Through Spelman's influence, the Indians "fraughted his ship with corn."

Soon after June 28, 1613, Argall sailed from Virginia on his "fishing voyage" in a well-armoured English man-of-war. His object was the French colony of Jesuits at Mt. Desert, now in Maine, but at that time within the bounds of Virginia. He attacked the buildings and returned with the priests late in July. He was sent back by Gates to destroy the buildings and fortifications there and at St. Croix and Port Royal. This was done and he arrived back at Jamestown, about the first of December. On this voyage, he stopped at New Netherlands, on the Hudson, and forced the colonists there to submit to the crown of England.

Shipbuilding on Plantations

The tracts of land or plantations occupied by individual settlers of the colony were very few until after the "starving time" in 1610. When the colony had been reorganized by Lord De La Warr and Sir Thomas Gates, and something like peace existed with the Indians, more land patents were issued year after year. A list of land owners, in 1625, in the records of the Company, shows nearly two hundred persons owning plots of land varying in size from forty acres to the thirty-seven hundred acres of Sir George Yeardley's plantation at Hungar's river on the Eastern Shore.

In *A Perfect Description of Virginia* by an unnamed writer in 1648, it is stated that there were in the colony "pinnaces, barks, great and small boats many hundreds, for most of their plantations stand upon the rivers' sides and up little creeks and but a small way into the land." Every planter must have had a boat of some kind. Neighborly communication had to be maintained, religious services attended, fishing and oystering to be done, crops of tobacco transferred to the ships anchored out in the channel, and cargoes of goods taken from the ships to the warehouses.

12

The planter navigated the boat himself unless he could provide a slave or an indentured servant.

Most of the shipbuilding done on the plantations was done by ship carpenters or men trained by them. The shipyards were very simple affairs, the essentials being a plot of ground on the bank of a stream with water deep enough to float the vessel and near a supply of suitable timber. Later would be added, perhaps, a small pier to which the boat could be be attached, and a small building or shed for the protection of tools.

A visiting ship in need of repair would seek some convenient place on the river and the hospitality of the neighboring planter. An instance is that of Captain Thomas Dermer from Monhegan, North Virginia, now in Maine, who arrived at the colony in September, 1619, in an open pinnace of five tons. He had met Captain Ward several weeks earlier at a place called "St. James his Isles," and there had put most of his provisions on board the *Sampson*, Captain Ward's boat. Of his arrival in Virginia, he wrote to Samuel Purchas as follows: "After a little refreshing, we recovered up the river to James Citie and from thence to Captain Ward his plantation, where immediately we fell to hewing boards for a close deck." He and his men soon fell sick with malaria and "were sore shaken with burning fever." As their recovery was slow and winter had overtaken them, Dermer decided to wait until spring before sailing north. Captain John Ward had arrived in Virginia during the previous April and was already a member of the House of Burgesses.

Some of the visitors did their shipbuilding more quickly. A Captain Thomas Young arrived in the colony with two ships on July 3, 1634, and by July 14, was reported by Governor Harvey to have built two pinnaces, and that he would be gone in two more days.

Some planters on the larger plantations continued to build their own ships even after public shipyards had been established in seaport towns. Flowerdieu Hundred on the James River was a prosperous plantation, where many vessels were built. It had its own wharf where large ships could be moored for loading.

Some shipbuilding at Westover on the James River is recorded in the diary of William Byrd II, who, after the death of his father in 1704, became owner of the plantation.

In July 1709, Byrd wrote: "I sent the boatmaker to Falling Creek to build me a little boat for my sea sloop." Two days later he wrote: "I sent Tom to Williamsburg for John B-r-d to work on my sloop." Later in the month, he noted that John B-r-d had come in the night to work on his sloop. In November, he wrote: "In the afternoon we paid a visit to Mr. Hamilton who lives across the creek. We walked about his plantation and saw a pretty shallop he was building." In August, 1710, he wrote that he had taken a walk to see the boatbuilder at work. On August 9, he wrote that he had paid the builder of his sloop sixty pounds, which was twenty pounds more than he had agreed for. Later in the year, he noted that his sloop had gone down to the ship-yard at Swinyards.

Byrd acquired a new shipwright who came from England on the ship *Betty* in 1711. In March, he wrote that the new ship-wright was offended because he had been given corn pone in-stead of English bread for breakfast. He had taken his horse and ridden away without a word. However, he reported later that the shipwright had returned. On May 15, 1712, Byrd reported that he had engaged Mr. T-r-t-n to build him a sloop next year. Sev-eral years later, he recorded the loss of his great flat boat, but it was found by a man at Swinyards. Swinyards was a place for public warehouses and a shipyard, located on the north bank of the James River, a short distance below Westover, opposite Windmill Point.

At Berkeley, a neighboring plantation on the James River, owned by Benjamin Harrison, there were extensive merchant mills and a large shipyard where vessels were built for the planta-tion. On October 20, 1768, there appeared a for-sale advertise-

14

ment in the *Virginia Gazette*: "A double decked vessel of 110 tons on the stocks at Berkeley Shipyard, built to carry a great burden, and esteemed a very fine vessel." Two years later, John Hatley Norton and a Mr. Coutts were negotiating with Colonel Harrison for the purchase of the ship *Botetourt* built there for which they offered 1100 pounds sterling. "She is as stout a ship as was ever built in America, and we expect will carry 380 hogsheads of tobacco," wrote Mr. Norton.

The Virginia Company's Interest in Boatbuilding

When Sir Thomas Smith ended his term as Treasurer of the Company in 1619, among many other charges brought against him by the opposing faction, it was declared there was left only one old frigate belonging to Somers' Isles, one shallop, one ship's boat, and two small boats belonging to private persons. In his defense, Smith referred to the 150 men he had sent to Virginia to set up iron works; the making of cordage, pitch, tar, pot and soap-ashes from material at hand; the cutting of timber and masts; and how he had sent men to erect sawmills for cutting planks for building houses and ships. In justification of Smith and himself, Robert Johnson, alderman, a leader during Smith's administration, drew up an account in which he stated among other evidences of prosperity that barks, pinnaces, shallops, barges, and other boats had been built in the colony; but this statement was not accepted as fact.

Sir Edwin Sandys succeeded Smith as Treasurer; and in the Earl of Southampton's administration in 1621, a list of improvements was drawn up, among which it was claimed that the number of boats was ten times multiplied and that there were four ships owned by the colony. A reply to this may be taken from *An Answer to a Declaration of the Present State of Virginia in May, 1623*, in which it was declared that the new administration was many degrees behind the old government, for in those times there were built boats of all sorts, barges, pinnaces, frigates, hoyes, shallops and the like.

The great massacre in March, 1622, put an immediate check on any progress in boatbuilding in the colony. For a time the settlers were panic stricken, and there was much talk of assembling all the remaining settlers on the Eastern Shore, but happily, wiser counsel prevailed. That the few boatwrights then in the colony perished is considered probable from the fact that none could be found to repair a boat that had drifted ashore at Elizabeth City after the massacre. When writing about the Indian massacre, Captain John Smith, in his *General History of Virginia*, in a bitter outburst, said: "Yea, they borrowed our boats to transport themselves over the river to consult on the develish murder that insued and of our utter extirpation." In Sir Francis Wyatt's commission to Sir George Yeardley on September 10, 1622, to attack the Indians in punishment for the massacre, he ordered the use of "such ships, barks, and boats as are now riding in this river as transports." The ships and barks may well have been English vessels.

When Virginia became a Crown Colony in 1624, the reports on the state of the colony named thirty-eight boats, two shallops, one bark, one skiff, and one canoe, but this was considered inaccurate as many plantations did not report their vessels.

Shipwrights and Ship-carpenters

Every colonizing expedition to the New World had been deeply impressed by the wealth of shipbuilding materials to be found. The English were particularly enthusiastic, since the scarcity of timber in England was very serious. Here, in Virginia, were to be found all that was needed for building ships: "oakes there are as faire, straight and tall and as good timber as any can be found, a great store, in some places very great. Walnut trees very many, excellent faire timber above four-score foot, straight without a bough." The report went on in praise of the tall pine

16

trees fit for the tallest masts, and the kinds of woods for making small boats: mulberry, sassafras, and cedar. Other materials were not wanting: iron ore, pitch, tar, rosin, and flax for making rope. The colonists saw in this wealth of materials a new source of supply at one-half of the previous cost. Both England and Holland had been purchasing their shipbuilding materials from Poland and Prussia at a cost of a million pounds sterling annually. One enthusiastic Englishman, when he heard these reports, wrote: "We shall fell our timber, saw our planks, and quickly make good shipping there, and shall return thence with good employment, an hundred sayle of ships yearly."

When Captain Newport returned to England in June, 1607, he carried with him a request, from the colonists to the company, for carpenters to build houses, and shipwrights to build boats. Upon Newport's return in 1608, he brought with him a number of Poles and Dutchmen to erect sawmills for the production of boards for houses and boats. This did not prove to be a successful venture. Further attempts were made in 1619, and later, to establish sawmills in the colony. Instructions sent to Governor Wyatt, in 1621, bade him "to take care of the Dutch sent to build sawmills, and seat them at the Falls, that they may bring their timber by the current of the water." Repeated appeals had been made to the Company for ship-carpenters without success. In January, 1621, the Governor and Council joined in an appeal for workmen to build vessels, of various kinds, for the use of the people in making discoveries, in trading with their neighbors, and in transporting themselves and goods from one place to another. In reply, a letter from the Company, in August, gave the encouraging news, that in the spring, the Company would send an excellent shipwright with thirty or forty carpenters. In preparation, they were advised to fell a large number of black oak trees, and bark as many others. The Company expected the sawmill to provide the planks and suggested a place near the sawmill and ironworks for the shipyard. A thousand pounds had been underwritten by private persons for sending the shipwrights and carpenters who were promised by the end of April at the latest.

The next spring, in May, the Council received notice that sailing on the ship *Abigail* were Captain Thomas Barwick and twenty-five other persons for building boats, ships and pinnaces. They were to be established together in an area of at least twelve hundred acres, and were to be employed only in the trade for which they were sent. Four of the Company's oxen were to be assigned to them for use in hauling the timber.

Captain Barwick and his men settled in Jamestown. At first they were employed in building houses for themselves and afterward began to build shallops, the most convenient and satisfactory vessels, for transporting tobacco to the large ships. Soon several of the men were ill, from malaria it was thought, and by the end of the year many of them had died. A letter from George Sandys, in March, to Deputy Treasurer Ferrar, sent by the ship *Hopewell*, told the discouraging news. He deplored the failure of the shipbuilding project caused by the death of Captain Barwick and many of his shipbuilders, "wherein if you blame us, you must blame the hand of God." He attributed the pestilent fever that raged in the colony to the infected people that came over in the *Abigail*, "who were poisened with stinking beer, all falling sick and many dying, everywhere dispersing the contagion." Not only the shipbuilders, but almost all the passengers of the *Abigail*, died immediately, upon their landing. The contagion even spread to the cattle and other domestic animals, it was said.

On March 31, 1626, Thomas Munn (?) came before the Council and the General Court of Virginia and swore that he was at the making of a small shallop, by direction of Captain Barwick, and that afterward this boat was sold to Captain William Eppes, for two hundred pounds of tobacco, and "as yet the debt is not satisfied unto any man." Upon the death of Captain Barwick, Munn had delivered to George Sandys, Treasurer, a list of debts owing, and this debt had never been paid.

Adam Dixon, who came over in the *Margaret and John*, was sent by the Company as a master calker of ships and boats. He was living at Pashbehays, near Jamestown, in 1624. As the years went by, a number of shipwrights came to the colony from time to time, and were engaged in private shipyards on plantations, or set up shipyards of their own. Orphan boys were sometimes apprenticed to these shipbuilders until they reached the age of twenty-one. They were expected to be taught to read, write and cipher in addition to learning the trade of ship-carpenter.

Many of the shipwrights who came to Virginia in the seventeenth century, became land owners, some of them owning large tracts of land, as shown by county records, especially in the Tidewater area. In Lancaster on the Rappahannock River, John Meredith, a shipwright, obtained, by patent, a tract of fifty acres. His sale of 600 acres is recorded, also a contract to build a sloop and a small boat, in payment of a debt of 47,300 pounds of tobacco.

In Rappahannock County records, we find shipwright Simon Miller, a noted shipbuilder, who owned a tract of 125 acres; and John Griffin, a shipwright, who, in 1684, recorded a deed to Colonel Cadwalader Jones for a bark of fifty odd tons, for the consideration of fifty pounds sterling.

The first John Madison of Virginia, great-great-grandfather of President James Madison, acquired considerable land in Virginia by the importation of immigrants; in a land patent dated 1682, he called himself a ship-carpenter. At this time, good ships of three hundred tons and over were being built in Virginia, and probably John Madison aided in the construction of one or more of these. It is evident that many of the shipwrights, who came to Virginia from England, found the life of a planter more desirable than that of a shipbuilder, while some of them combined the two occupations.

Controversies Over Boats

The Council and General Court of Virginia were called upon occasionally to settle controversies over vessels of various kinds and to hear reports concerning others. The following reports are from the records of the Court for 1622 to 1632.

At an early date, Robert Poole reported a trading voyage with the Indians for Mr. "Treasurer," in the pinnace *Elizabeth*, during which he gave ten arms length of blue beads for one tub of corn and over, and thirteen arms length for another tub. Anne Cooper complained that her late husband, Thomas Harrison, loaned a shallop to Lieutenant George Harrison, late deceased. It was ordered by the Court that she should receive one hundred pounds of merchantable tobacco from George Harrison's estate.

An argument between John Utie and Bryan Caught resulted in the order that the latter should build Utie a shallop eighteen feet, six inches keel; six feet, six inches breadth; with masts, oars, yard and rudder, and to find the 1100 nails and six score "ruff and clench" desired. Utie was to pay Bryan for building the shallop six score pound weight of tobacco, and to furnish the help of a boy and the boy's diet. Also, he was to pay Bryan six score pounds of tobacco for a boat previously built for him.

Captain Francis West, a member of the Council, desired that he be given the use of the Spanish frigate with all her tackle, apparel, munitions, masts, sails, yards, etc., that had been captured by John Powell, with a shallop built for that purpose, on an expedition to the West Indies in the man-of-war, *Black Bess*. He was required to pay 1200 pounds of tobacco to the captain and men.

In trading for corn for Southampton Hundred, John Powntis was allowed a barrel of the corn for the use of his pinnace.

Mr. Proctor had to pay Mr. Perry fifty pounds of tobacco for splitting Perry's shallop. Later, a shallop, which Edmund Barker sold to Mr. Rastall's men, was ordered returned to Mr. Perry, and Edmund Barker to be paid fifty pounds of tobacco for mending the shallop. To settle a charge against Thomas Westone by several men, he was ordered to appear before the Governor with his pinnace. At a later meeting, Thomas Ramshee swore that Westone was owner of the ship *Sparrow* and "did set her out of his own charge, from London to Virginia." This was an early seagoing vessel of a colonist, but whether built in Virginia, or purchased, is not stated.

Nicholas Weasell received the most severe penalty, in cases concerning boats, when he was ordered to serve Henry Geny the rest of the year from February, for taking away Geny's boat without leave, "whereupon it was bilged and spoiled."

Captain Claiborne purchased a shallop with appurtenances from Captain John Wilcox who had been "at the plantation called Accomack" since 1621. He paid Wilcox 400 pounds of tobacco for the shallop, and sold it to Thomas Harwood. Captain Wilcox failed to make delivery, and the court ordered the attorney of Captain Wilcox to make satisfaction to Thomas Harwood.

The court was called upon to settle a controversy between Captain William Tucker and Mr. Roland Graine about a boat. A Mrs. Hurte was named as the owner of another ship in the colony, the *Truelove*, formerly owned by John Cross, deceased in England. A much discussed case was that of William Bentley, on trial for the killing of Thomas Godby, which resulted when Mr. Conge's boat ran ashore at Merry Point, near William Parker's house. While there, Bentley, who had arrived in the boat, got into a quarrel and fight with Godby, and was accused of killing him.

These Court records show that most of the cases concerned vessels built in the colony: boats, pinnaces, and shallops. The ships mentioned were evidently of English make. The shallop was the most popular boat for use in the colony. It was a small boat from sixteen to twenty feet in length, fitted with one or two masts and oars, and suitable for exploring the creeks and rivers, collecting corn from the Indians, and transporting tobacco to waiting ships.

The Eastern Shore records are among the earliest in Virginia. Shipbuilding in the early days has been ably discussed by Dr. Susie M. Ames in *Studies of the Virginia Eastern Shore in the Seventeenth Century*. In 1630, John Toulson, or Poulson, built a pinnace at Nassawadox in which he had one-half interest. Richard Newport, one of Captain Christopher Newport's sons, while living in Northampton County, bought a shallop from the carpenter, Thomas Savage, for the use of the merchant, Henry Brookes, for which Savage was paid twenty pounds sterling. William Berry, another Eastern Shore carpenter, made an agreement with Philip Taylor, one of William Claiborne's men, during the Kent Island controversy, to make him a boat, twenty by ten feet, provided Taylor furnished the boards for the deck between the forecastle and the cabin. For this, Berry was to receive two cows with calf and four hundred pounds of tobacco. During the dispute over Kent Island, a pinnace, belonging to Captain Claiborne, was taken by the Marylanders.

Obedience Robins, a well-known citizen of the Eastern Shore, acquired from the boatwright, William Stevens, a shallop, twenty-six feet in length, with masts, yards, and oars. He owned a pinnace also, which he had named *Accomack*. A number of lawsuits on the Eastern Shore in the 1640's, involved boats and ship materials. Philip Taylor was indebted to William Stevens for one house, four days on a shallop, valued at one pound sterling, six gallons of tar, and 1250 nails of various sizes. Payment was ordered made to the overseers of the estate of Daniel Cugley of one small boat, twenty-four yards of canvas, twenty gallons of tar, and ninety ten-grote nails, supplies for making a boat. Another court order concerned the delivery of a boat, and 3500 six-penny nails lent by John Neale. Ambrose Dixon testified that he and his mate had built a boat for Randall Revell. In 1638, two planters of Accomack, Nicholas White and one Barnaby, made voyages to New England in their own vessels. The names of Walter Price and Christopher Stribling shipwrights are listed in the early records of Northampton County.

Encouragement for the Building of Ships

The General Assembly of Virginia encouraged shipbuilding by such laws as those enacted during 1662: "Be it enacted that every one that shall build a small vessel with a deck be allowed, if above twenty and under fifty tons, fifty pounds of tobacco per ton; if above fifty and under one hundred tons, one hundred pounds of tobacco per ton; if above one hundred tons, two hundred pounds per ton. Provided the vessel is not sold except to an inhabitant of this country in three years."

Other encouragement by Virginia to owners of vessels, built by them, was the exemption of the two shillings export duties per hogshead of tobacco; the exemption from castle duties; the reduction to two pence per gallon on imported liquor from the four pence required of foreign vessels; and the exemption from duties imposed on shipmasters on entering and clearing, and for licenses and bond where necessary.

The English government discouraged manufacture in the colonies that would compete with home manufactures, but the building of ships was an exception. England needed ships and granted the colonies the right to build as many as they could. Throughout the whole period of royal government, there were enacted various laws remitting the duties on imports brought in on native ships and remission of tonnage duties. This aroused the resentment of the English shipbuilders, who had endeavored to put a stop to the building of ships of any size in the colonies. They were alarmed, too, at the laws passed in the colonies to encourage shipbuilding and complained that they had been discriminated against. Resolutions were passed by Parliament to investigate such laws framed in the colonies, and a bill, based upon these resolutions was proposed, but never introduced.

However, in 1680, Governor Culpeper was ordered to annul the laws exempting Virginia owners of vessels constructed in the colony from duties on exported tobacco and castle duties. The grounds upon which this order was based were (1) the injustice of granting privileges to Virginia ship owners, not enjoyed by the owners of English vessels, trading in Virginia waters; (2) the success of the navigation laws would be impaired by creating a Virginia fleet, able to transport tobacco, without the assistance of English vessels; and (3) owners of English ships might be tempted to order them as belonging to Virginians. Since the Virginia fleet in 1681, was composed of two ships, as mentioned by John Page, in a petition to Lord Culpeper, the English were thought to be unnecessarily alarmed.

During the 1660's, following the laws of the General Assembly, a number of Virginia built ships were recorded. There was much shipbuilding activity on the Eastern Shore. The mate of the *Royal Oake*, when caught trading illegally, stated that the owner had another boat in the house of a Mr. Waters, and also had a sloop being built there. About this time, a shipwright agreed to build between May and October, for William Whittington, a sloop of twenty-six feet keel, and breadth in proportion, receiving for his work 4,400 pounds of tobacco. In 1666, John Goddon entered a claim for a vessel of twenty-five tons built for him in Accomack. John Bowdoin built a brigantine which he named *Northampton*.

The size of the vessels built in Virginia had been increasing steadily. Thomas Ludwell, Secretary of the Colony, reported, in 1655, that there had been built recently, several small vessels which could make voyages along the coast, presumably sloops. Again, in a letter to Lord Arlington, Secretary Ludwell made the following statement: "We have built several vessels to trade with our neighbors, and do hope ere long to build bigger ships and such as may trade with England."

Colonel Cuthbert Potter of Lancaster County, who was sent on

a mission to ascertain the truth of the reported Indian depredations in Massachusetts and New York, was an early settler in
the colony, and had acquired large land holdings in Middlesex
County. About 1660, he removed to Barbadoes in his own sloop,
the *Hopewell*.

In 1665, James Fookes agreed to build for the widow, Mrs.
Ann Hack, a sloop that would carry thirty-five hogsheads of tobacco, if Mrs. Hack would supply the plank and a barrel of tar;
Fookes agreed to finish the job by the 25th of December. The
following summer, at the plantation of Mrs. Hack, Fookes made
a formal contract with the brother of Mrs. Hack, Augustine Herrman of Bohemia Manor in Maryland, to build a sloop and have
it ready by the following October. Herrman is well-known for
his 1673 map of Maryland and Virginia. Twenty years later, the
dimensions of the *Phenix*, another vessel built by Fookes, were
given: length of keel, forty feet; breadth, fourteen feet, nine
inches inside; depth, eight feet, ten inches.

In the English *News Letter* of March 12, 1666, was carried
an encouraging news item: "A frigate of between thirty and forty
[tuns?], built in Virginia, looks so fair, it is believed that in a short
time, they will get the art of building as good frigates as there
are in England." At that time, a new fort was being erected at
Point Comfort, and it was ordered that every ship riding in the
James River should send one carpenter with provisions and tools
to work on this fort.

In 1667, Mrs. Sarah Whitby, widow of John Whitby, petitioned the King in Council as follows: "The petitioner with other
planters in Virginia are owners of the ship *America*, built in
Virginia by Captain Whitby, and pray for a license, for the said
vessel with six mariners, to proceed to Virginia." The workmanship of the *America* and her fine appearance had aroused the
interest of the English, and expectations arose that Virginia might
soon become skillful in building large vessels.

In a reply by Sir William Berkeley, Governor of Virginia, to an inquiry by the Lords Commissioners of Foreign Plantations, in 1671, as to the number of ships that trade yearly with the colony, he answered that there were a number of ships from England and Ireland and a few ketches from New England, but never at one time more than two Virginia-owned vessels, and they not more than twenty tons burden. He stated further that the severe Act of Parliament which excluded the colony from commerce with any other nation, was the reason why "no small or great vessels are built here." But other records of the time contradict Berkeley's statement as to the number and size of vessels built in the colony. In addition to those mentioned above, there is found in the records of York County, an itemized cost of building a sloop, the total amount being 4,467 pounds of tobacco. The various materials were furnished by the owners: Richard Meakins, 950 feet of plank; Mr. Newell, the rigging; Captain Sheppard, the sail; and Mr. Williams, the rudder iron. About four months were required to complete the vessel, charges for food running that length of time, during which a cask of cider was consumed. Some sloops were made large enough to hold as many as fifty hogsheads of tobacco, and could sail outside the coast. The sloop *Amy*, with fourteen hogsheads of tobacco, sailed from Virginia to London in 1690.

Dr. Lyon G. Tyler in *The Cradle of the Republic* wrote that as early as 1690, ships of 300 tons were built in Virginia, and trade in the West Indies was conducted in small sloops. Lieutenant John West of the Eastern Shore, stating that he had built a vessel of forty-five tons, decked and fitted for sea, petitioned the court for a certificate to the Assembly as encouragement for so doing. Two other shipwrights, Thomas Fookes and Robert Norton, testified as to the weight of the vessel. West was evidently seeking the subsidy of fifty pounds of tobacco for building a vessel "above twenty and under fifty tons," under the law of 1662.

John West was evidently considered an excellent boatwright

and carpenter, for in an indenture of the year 1697, made between him and Robert Glendall, late of Elizabeth City County, West is enjoined by the court to do his untmost to 'instruct Glendall in sloop and boat building, and in such other carpenter's work as he was "knowing in."

In his testimony before the Board of Trade on September 1, 1697, as to the manufactures in Virginia, Major Wilson stated that very good ships were built in Virginia of 300 tons and upwards; but cordage, iron, and smith's work were "brought thither." During that year, a group of merchants in Bristol, England, had a number of ships constructed in Virginia. They were influenced by the fine quality of timber and the small cost of the work, as compared with the cost of similar work in England. Also, a matter of no small importance, a cargo of tobacco was ready for each completed ship.

The wills of deceased persons sometimes revealed ownership of vessels. Of particular interest is the will of Nathaniel Bacon, Senior, in which he left to his wife and his nephew, Lewis Burwell, "all ships or parts of ships . . . to me belonging in any part of the world." These were to be disposed of by Abigail, his wife, and the nephew as they saw fit. An inventory of the estate of one Thomas Lloyd of Richmond County, on October 27, 1699, lists one decked sloop on the stocks, unfinished, of about thirty tons; one small open sloop newly launched, not finished, of twenty-five tons; one new flat, one old ditto; one old barge; one parcel of handsaws, etc.

Sir Edmund Andros, Governor of Virginia, in answering the inquiries of the Council of Trade and Plantations, the clearing house for colonial affairs, in the year 1698, stated that there were 70,000 inhabitants in Virginia, and the number of vessels reported by the owners were four ships, two barks, four brigantines, and seventeen sloops. His report for the previous year had named eight ships, eleven brigantines, and fifteen sloops that had been built for which carpenters, iron work, rigging, and sails had been brought from England.

EIGHTEENTH CENTURY SHIPBUILDING

The building of ships, barkentines and sloops in Virginia, during the early years of the eighteenth century, had so increased that the Master Shipbuilders of the River Thames addressed a petition to the King in 1724, stating that by the great number of ships and other vessels lately built, then building, and likely to be built in the colonies, the trade of the petitioners was very much decayed, and great numbers of them for want of work to maintain their families, had of necessity left their native country and gone to America. They felt that not only British trade and navigation had suffered thereby, but danger existed in fitting out the Royal Navy in any extraordinary emergency. This petition applied to the northern colonies particularly, as they were far ahead of Virginia in shipbuilding, but the southern colonies were included. As we have seen, many shipwrights came to Virginia and acquired large tracts of land and became planters.

In the narrative of his travels in Virginia, with some companions early in the eighteenth century, Francis Louis Michel of Berne, Switzerland, related that when he was within fifty miles of the coast, he saw two ships, the larger, one of the most beautiful merchantmen he had ever seen. Because it was built in Virginia, it was named *Indian King* or *Wild King*, he did not remember which. Three years before, it had fallen into the hands of pirates, so the narrative related, but had been rescued by the British warship *Shoreham*, and sixty pirates of all nations taken prisoners, all of whom were hanged in England.

How many vessels were built or repaired at the Point Comfort shipyard is not known. At a meeting of the Council of Virginia in May, 1702, a letter from Captain Moodie stated that he had fitted up a very convenient place at Point Comfort for careening Her Majesty's ships of war, or any other ships that came to the colony; and he proposed that some care be taken and some person appointed to have charge of the situation. This arrangement was confirmed by a letter from Lieutenant Governor Alexander Spotswood to the British Admiralty on October 24, 1710, in which he wrote that for the convenience of careening, there is a place at Point Comfort which, with a small charge, could be fitted up for that purpose; H.M.S. *Southampton* had careened there, and there may be served the largest ships of war, which Her Majesty will have occasion to send to Virginia as cruisers or convoys. This careening site at Point Comfort provided long-needed facilities for careening vessels for repairs and scraping bottoms. As early as 1633, David Pietersz de Vries from Holland, arrived at Jamestown with a leaky ship, but found no facilities in the colony for careening vessels. He found it necessary to sail to New Netherlands for such repairs. As late as 1700, when the *Shoreham*, a fifth rate frigate, was the Chesapeake Bay guardship, Captain Passenger, her commander, wrote to Governor Nicholson: "I have only to offer. (may your Excellency think convenient) about the latter end of September to careen the *Shoreham*. She is at present very foul, and the rudder is loose, which I fear before the next summer, may be of dangerous consequences which cannot be removed, without careening or lying ashore, which I presume there is no place in Virginia, that will admit of." It is thought, however, that there must have been careening places in the colony for the smaller vessels, or how else could the pinnaces and sloops have been kept in repair.

Sloops became popular in the eighteenth century, and a number of them were built in Virginia to be disposed of in the West Indies. After the sloop was finished, she received a cargo of tobacco, and vessel and tobacco were sold together. Because of the danger from pirates and Spanish interference, the sloops for the West Indies trade were designed especially for speed and maneuverability. The pilot boat evolved in the colony quite early. An advertisement appeared in the *Virginia Gazette,* on July 22, 1737,

for a pilot boat stolen or gone adrift from York River. The boat
was twenty-four feet keel, nine feet beam, with two masts and
sails, and was painted red. Another advertisement in September,
1739, concerning a boat stolen from Newport News, on the James
River, by one James Hobbs, a carpenter. The boat was about
fifteen feet keel, had two masts, and was payed with pitch. It
had a new arch thort of black walnut, and a tarpaulin upon the
forecastle.

Norfolk became one of the busiest ports in Virginia, both in
shipbuilding and ship repair work. A shipyard had been established on the Elizabeth River in 1621 by John Wood and work
had been almost continuous, though at times very slow, throughout the seventeenth century. An inventory in 1723, listed one
brigantine, three sloops, and three flats owned by Robert Tucker.
One of the sloops was forty feet in length and valued at 230
pounds sterling. Captain Samuel Tatum owned the ship *Caesar*,
which was said to be worth 625 pounds sterling, and the sloop
Indian Creek valued at twenty-five pounds. William Byrd in
his *History of the Dividing Line*, states that he saw at Norfolk,
in 1728, twenty sloops and brigantines. Some of them were quite
evidently of English origin. In 1736, the sloop *Industry*, "lately
built in Norfolk," was loaded with tobacco in the James River
to take to London.

Captain Goodrich, master of the ship *Betty* of Liverpool, which
was built on the Elizabeth River for the Maryland trade, was
permitted by the Council of Virginia, to sail to Liverpool without the payment of the usual port duties. The firm of John
Glasford and Company contracted with Smith Sparrows in 1761,
for a ship built at Norfolk, sixty feet in length, sixteen feet in
the lower hold, and four feet between decks, the price being
fifty shillings per ton.

Many of the shipwrights, who came to Virginia and became
land owners, settled in Norfolk. That port was especially known
for this kind of citizen, ranking next to the merchant in wealth

and influence. Among house owners were some ship-carpenters who carried on their trade, receiving for a day's work four shillings and a pint of rum, more wages than the salary of some clergymen. Several shipwrights listed in Lower Norfolk were large property owners. Abraham Elliott owned land both in Virginia and England. One John Ealfridge owned one-half interest in a mill, and acquired a plantation for each of his two sons in addition to his own.

To secure a large sum of money due Robert Cary of London, Theophilus Pugh of Nansemond County mortgaged his lands, slaves, and vessels with all their boats. The vessels were listed as follows: ships, *William and Betty, Prosperous Esther;* sloops, *Little Molly, Little Betty;* schooners, *Nansemond Frigate, Pugh.* If the average planter had owned the equivalent of two ships, two sloops and two schooners, the total number of vessels in Virginia in the middle of the eighteenth century would have far exceeded any inventory reported.

The frame of a snow, which was to have been built by Thomas Rawlings, a ship-carpenter, for Mr. John Hood, merchant of Prince George County, was advertised for sale in 1745. The snow was to have been sixty feet keel; twenty-three feet, eight inches beam; ten feet hold; and four feet between decks. Also advertised for sale about the same time was a schooner, trimmed and well-fitted with sails and rigging to carry fifty hogsheads of tobacco. In March, 1746, the sloop *Little Betty,* burden fifty tons, was offered for sale with her sails, anchors, furniture, and tackle.

The advertisements of Virginia-built vessels in the 1750's, and in the 1760's, show a steady increase in the size of sloops and ships. The following are mentioned: a brig of eighty tons; several snows, one to carry 250 hogsheads of tobacco; and several schooners. Schooner rigged boats appeared in the colony early in the eighteenth century, and gradually increased in size and importance. During the second half of the eighteenth century, the schooner displaced the sloop as the principal coastwise vessel, and emerged during the Revolution as a distinctive American type. "The most spectacular event in the history of naval architecture in the eighteenth century was the emergence of the Chesapeake Bay clipper-schooner," says Arthur Pierce Middleton.

In April, 1767, John Hatley Norton came from London to be his father's agent with headquarters in Yorktown. He wrote home that his cousins, the Walker Brothers, had a shipyard at Hampton, and were building ships of new white oak, well calculated for the West Indies trade. A letter from John M. Jordon & Company, London, in 1770, reads in part as follows: "Mr. William Acrill desires you will make insurance of his brig, *America,* Captain William C. Latimer; in case of loss to receive four hundred pounds. She is chartered by a gentleman on the Rappahannock; and is now in Hampton Roads, and will sail tomorrow or next day; and in case she arrives safe, you are to receive her freight, and sell the vessel, provided you can get four hundred pounds for her."

Occasionally, we find an account of the use of a vessel of some kind or other for pleasure. In Fithian's *Journal and Letters,* the author writes in 1773, that his employer, Mr. Robert Carter of Nomini, prepared for a voyage in his schooner *Harriot* (named for his daughter), to the Eastern Shore of Maryland for oysters. The schooner was of forty tons burden, thirty-eight feet in length, fourteen feet beam, six feet in depth of hold, carried 1400 bushels of grain, and was valued at forty pounds sterling. Again from the *Journal:* "From Horn Point, we agreed to ride to one Mr. Camel's, who is Comptroller of the customs here. Before dinner, we borrowed the Comptroller's barge, which is an overgrown canoe, and diverted ourselves in the river which lies fronting his house."

Trading Towns and Ports

In the early days of the colony after tobacco had become a commodity for export, ships moored at the wharves of the plantations along the James, York and Rappahannock rivers and their

Photograph by W. T. Radcliffe.

Susan Constant. Replica of the Ship that brought the first settlers to Jamestown, 1607.

Photograph by W. T. Radcliffe.

Interior of the *Susan Constant*

The manner of makinge their boates. XII.

He manner of makinge their boates in Virginia is verye wonderfull. For wneras they want Inſtruments of yron , or other like vnto ours, yet they knowe howe to make them as handſomelye , to ſaile with whear they liſte in their Riuers, and to fiſhe with all, as ours. Firſt they chooſe ſome longe , and thicke tree, accordinge to the bignes of the boate which they would frame , and make a ſyre on the grownd abowt the Roote therof, kindlinge the ſame by little , and little with drie moſſe of trees , and chipps of woode that the flame ſhould not mounte opp to highe, and burne to muche of the lengte of the tree· When yt is almoſt burnt thorough, and readye to fall they make a new ſyre, which they ſuffer to burne vntill the tree fall of yt owne accord. Then burninge of the topp , and bowghs of the tree in ſuche wyſe that the bodie of theſame may Retayne his iuſt lengthe , they raiſe yt vppon potes laid ouer croſſ wiſe vppon forked poſts, at ſuche a reaſonable heighte as they may handſomlye worke vppó yt. Then take they of the barke with certayne ſhells: thy reſerue the, innermoſt parte of the lennke , for the nethermoſt parte of the boate. On the other ſide they make a ſyre accordinge to the lengthe of the bodye of the tree, ſauinge at both the endes. That which they thinke is ſufficientlye burned they quenche and ſcrape away with ſhells, and makinge a new ſyre they burne yt agayne, and ſoe they continne ſomtymes burninge and ſometymes ſcrapinge, vntill the boate haue ſufficient bothowmes. This god indueth thiſe ſauage people with ſufficient reaſon to make thinges neceſſarie to ſerue their turnes.

Indian Dugout Canoe

Rose's Tobacco Boat, 1749

Rucker's Tobacco Boat, 1771

From Percy's *Piedmont Apocalypse*.

Shallop

From a sketch by Gordon Grant.

Discovery. Replica of the pinnace that accompanied the *Susan Constant,* 1607

Photograph by W. T. Radcliffe.

Construction of the *Discovery*, after Seventeenth-Century Shipbuilding

From Abbot's *American Merchant Ships.*

An Early Shipyard

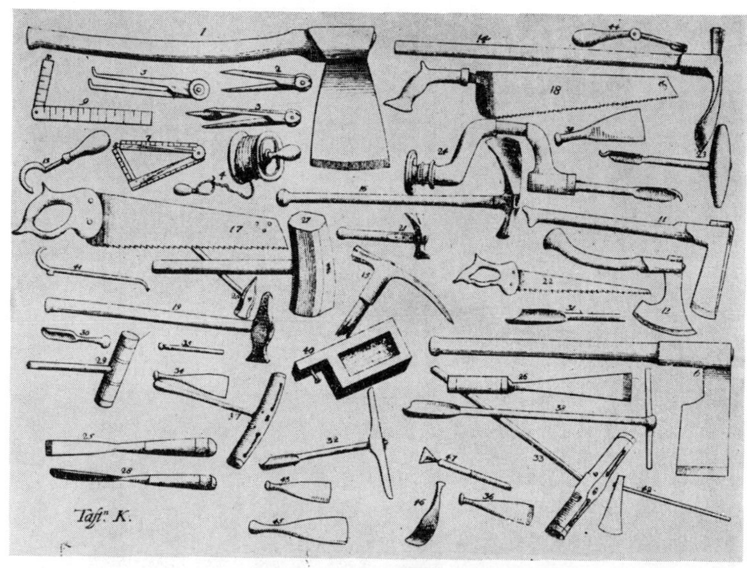

From Ralamb's *Skeps Byggerij*, 1691.
Trans. by J. Aasland, Jr., Hampton, Va.

Early Shipbuilding Tools used in Sweden and Other Countries

1—English Broad Axe. 2—Compass. 3—Compass with Chalk Holder. 4—Chalk Line on Roller. 5—Compass. 6—Axe for Holes. 7—Ruler. 8—Tongue on Ruler 1½ ft. 9—Dutch Ruler. 10—Tongue on Ruler for Ship layout. 11—Swedish Cutting Axe. 12—Trimming Hatchet. 13—Hook for removing old calking. 14—English Adz. 15—Adz. 16—Swedish or Dutch Adz. 17—English Handsaw. 18—Handsaw with Handle. 19—Mallet. 20—Hammer. 21—Claw Hammer. 22—Circle Saw. 23—Auger. 24—Dutch Brace Auger. 25—English Wood Chisel. 26—Wood Chisel. 27—English Mallet. 28—Gouge. 29—Swedish Mallet. 30—Gouge. 31—Gouge. 32—Gouge. 33—Calking Mallet. 34—Calking Tool. 35—Spike Iron. 36—Calking Tool. 37—Calking Mallet. 38—English Gouge. 39—Calking Iron. 40—Lubricating Tool, also for removing pitch. 41—Hook for removing oakum or old calking. 42—Calking Iron. 43—Calking Iron. 44—Tool used to clean out seams. 45—Calking Iron. 46—Calking Iron. 47—Scraper.

From Pepysian MSS in Magdalene College, Cambridge, England.

Shipwrights Drawing, 1586

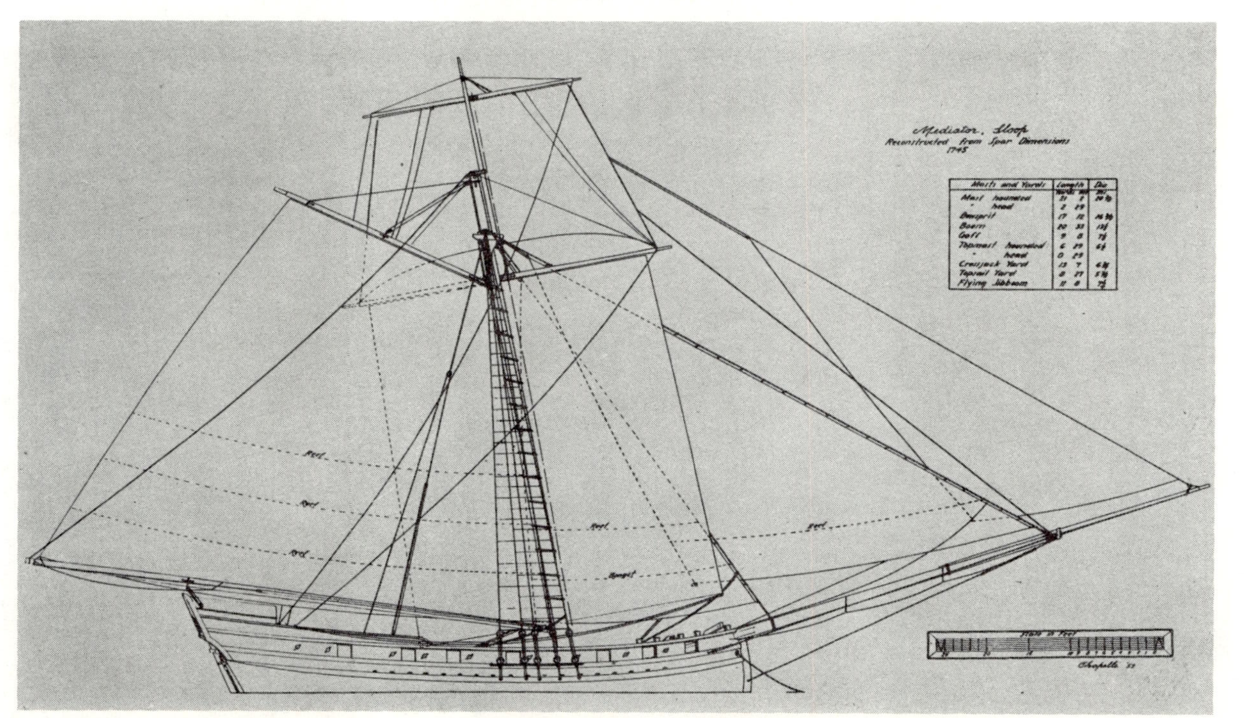

Drawn by H. I. Chapelle from Admiralty Records.

H.M.S. *Mediator*, a Virginia Sloop of about 1741, Purchased for the Royal Navy in 1745.

From an original drawing, 1755.

Sloops in the York River between Yorktown and Gloucester Point

Chesapeake Bay Log Canoe under construction

From *Naval Chronicle*, 1815.

A Virginia Pilot Boat with a view of Cape Henry

From a watercolor by G. Tobin in the National Maritime Museum, London.

American Schooner off Coast of Virginia, 1794

From a painting of Curacao, 1785.

British Schooner

Seventeenth-Century Shipyard in England

From the Science Museum, South Kensington, London.

From the Science Museum, South Kensington, London.

Careening Ships in England, 1675

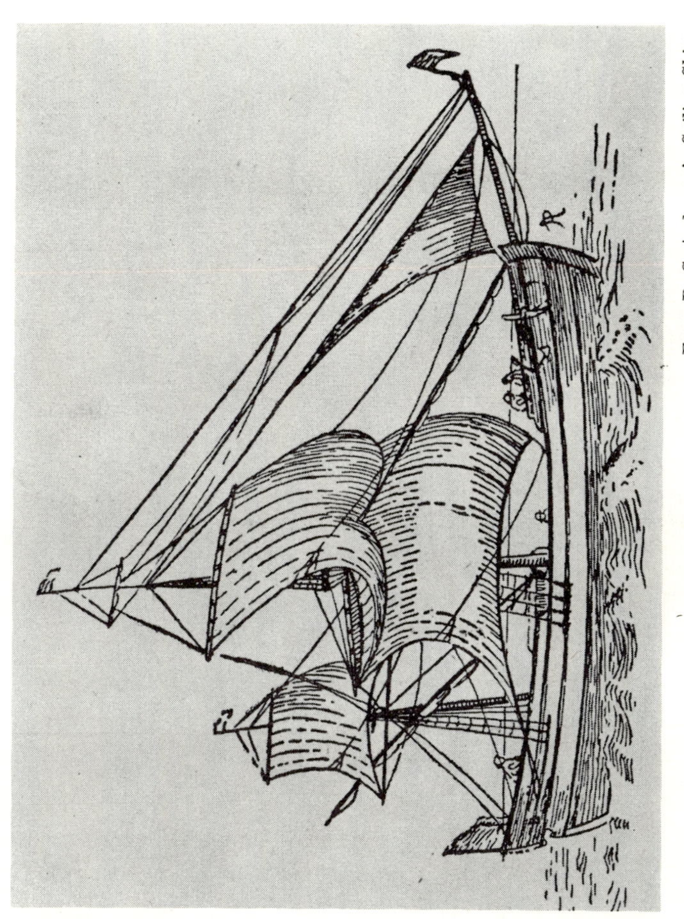

From R. C. Anderson's *Sailing Ships.*

English Ketch, about 1700

From Williams' *Sailing Vessels of the Eighteenth Century.*

Brigantine, about 1720

From Williams' *Sailing Vessels of the Eighteenth Century.*

Brig

From Williams' *Sailing Vessels of the Eighteenth Century.*

Snow

From the Archives in the Custom House, London.

Small Galley-built Vessel, Ship-rigged, 1714

Photograph by W. T. Radcliffe.

SS *United States*, Built at the Newport News Shipbuilding and Dry Dock Company. Latest shipbuilding in Virginia, to compare with Seventeenth-Century Craft.

estuaries. As trade increased, larger ships were used which anchored in the channels of the rivers, and the tobacco and other exports were carried to them by small boats—shallops, sloops, and barges. The government complained that it was losing revenue by this individualistic and unorganized shipping of the planters, and steps were taken to correct this. In 1633, it was enacted by the General Assembly that all goods entering in any vessel—ship, bark or brig, should discharge at Jamestown. This Act applied to the colonists in their exports as well, but the law was disregarded.

In 1680, places were selected in the different counties that had the advantage of accessibility and deep water where ships could gather to receive and discharge their cargoes. The establishment of these trading towns, as they were called, was by an Act as follows:

The General Assembly having taken into consideration the great necessity, usefulness and advantages of cohabitation . . . and considering the building of storehouses for the reception of all merchandizes imported, and receiving and laying ready all tobacco for exportation and sale . . . that there be in every respective county fifty acres of land purchased by each county and laid out for a town and storehouses . . .

The price of the fifty acres of land was set at 10,000 pounds of tobacco and casks. Lots of one-half acre were to be sold to individuals by a stated time at the price of one hundred pounds of tobacco. Twenty places were named in the counties where trading towns were to be established:

Henrico, at Varina. Charles City, at Flower de Hundred opposite Swinyards. Surry, at Smith's fort. James City, at James City. Isle of Wight, at Pate's Field, Pagan creek. Nansemond, at Huff's point. Warwick, at the mouth of Deep creek. Elizabeth City, west side of Hampton river. Lower Norfolk, on Nicholas Wise's land. York, on Mr. Reed's land. New Kent, at the Brick House. Gloucester, at Tindall's point. Middlesex, west side of Wormley's creek. Rappahannock, at Hobb's hole. Stafford, at Peace point. Westmoreland, at Nominy. Accomack, at Onancock. Northampton, north side of King's creek. Lancaster, north side of Corotomond creek. Northumberland, at Chickacone creek.

The towns were building up. Warehouses, churches, and prisons were erected in many of them, as well as private dwellings. An occasional court house could be found where legal proceedings were enacted. In 1691, however, an Act of the General Assembly changed many of the trading towns to ports, but was suspended later until the pleasure of the King and Queen on the subject should be learned. No definite action was taken until 1705, when Queen Anne, who ascended the throne in 1702, expressed approval. Then an Act for ports of entry and clearance was passed to be in use from the 25th of December, 1708. This Act provided that naval officers and collectors at the ports should charge Virginia owners of vessels no more than half of the fees required for the services of entering and clearing. The sixteen towns to become ports were named as follows:

Hampton. Norfolk. Nansemond. James City. Powhatan (Flower de Hundred). Yorktown. Queensborough, at Blackwater. Delaware, at West Point. Queenstown, at Corrotoman. Urbanna, at Middlesex. Tappahannock, at Hobb's hole. New Castle, at Wicomico. Kingsdale, at Yohocomoco. Marlborough, at Potomac creek. Northampton, at King's creek. Onancock.

The names of some of the trading towns were changed when they became ports, and soon became important and well-known throughout the country. Hampton, known first by the Indian name Kecoughtan (spelled in various ways) was settled in 1610. Although the name had been changed to Elizabeth City by the Company in May, 1620, upon the petition of the colonists, the old Indian name was still in use occasionally in the 18th century. In papers relating to the administration of Governor Nicholson is a list of vessels about to sail from "Keccowtan" in July 1705, sixty-seven sail of merchant ships bound for various ports of Great Britain. The names Kecoughtan, Elizabeth City, Lower James, and even Southampton were used interchangeably, and shown on records of the colony, until the Act of 1705, named the port Hampton. In British colonial records of 1700, we find Hampton Town, Elizabeth City and Keccowtan used in the same chapter.

F. C. Huntley in his *Seaborne Trade in Virginia in Mid-Eighteenth Century*, published in the *Virginia Magazine of History*, vol. 59, makes the statement that in the 18th century, Port Hampton handled the largest amount of shipping of all the Virginia ports, judging from the total tonnage of vessels entering and clearing as given in the records of the Naval Officers. He uses 1752, as a normal trade year of which he gives interesting statistics. He states that the tonnages that entered and cleared the Port Hampton naval office were distributed among five different types of rigging. Cleared: 64 sloops, 46 schooners, 16 ships, 20 brigs, 10 snows. Entered: 59 sloops, 40 schooners, 40 ships, 18 brigs, 12 snows. Of these a goodly portion were built in Virginia.

After taking part in laying the dividing line between Virginia and North Carolina, William Byrd II wrote on March 28, 1728:

Norfolk has most the air of a town of any in Virginia. There were more than 20 brigantines and sloops riding at the wharves and oftimes they have more. It has all the advantages of a situation requisite for trade and navigation. There is a secure harbor for a goodly number of ships of any burthen. The town is so near the sea that a vessel can sail in and out in a few hours. Their trade is chiefly to the West Indies whither they export abundance of beef, pork, flour and lumber.

In the *Journal* of Lord Adam Gordon, Colonel of the 66th Regiment of Foot, stationed at the West Indies from 1763 to 1775, is extracted the following: "Norfolk hath a depth of water for a 40-gun ship or more, and conveniences of every kind for heaving down and fitting out large vessels; also a very fine ropewalk. There is a passage boat from Hampton to Norfolk and from York to Gloucester." In the third quarter of the 18th century, Norfolk became the principal seaport of Virginia.

Yorktown was founded on land patented about 1635 by Nicholas Martiau, a Walloon who had come to Virginia in the summer of 1620. His grandson, Benjamin Read, sold fifty acres to the colony in 1691, and here Yorktown as a port built the first custom house, not only in Virginia, but in the country. A two-story brick building, erected about 1715, by Richard Ambler, who occupied the building as collector of customs for Yorktown in 1720. It became a port of entry for New York, Philadelphia and other northern cities, the importance of which was destroyed by the Revolutionary War. York County was one of the eight original shires in 1634, under the name, Charles river, changed in 1643 to York. The old custom house is still standing and is used as a museum for colonial and revolutionary relics.

The location of Alexandria on a large circular bay in the Potomac river soon gave that town great importance as a port and shipyard. For generations, tobacco and grain were shipped from there, and imports of many kinds brought in. Master shipbuilders turned out vessels manned, owned and operated by Alexandrians. From her ropewalk came the rope to hoist the sails made in her sail lofts. On May 19, 1760, George Washington went to Alexandria to see Col. Littledale's ship launched. He tells of another launching he attended there on October 6, 1768, when he "stayd up all night to a ball."

The two creeks flowing from near Williamsburg to York river on one side and the James on the other, played an important part in early colonial history. From York river sloops, schooners, barges and all manner of flat-bottomed craft sailed up Queen's creek to Queen Mary's port with its Capitol Landing within a mile of Williamsburg. The same kind of watercraft sailed from James river up College creek to Queen Anne's port with its College Landing near the city. Cargoes of mahogany, lignum vitae, lemons, rum, sugar and ivory were discharged. Received in return were tobacco, grain, flour and other commodities. Vessels on Queen's creek were required to pass through the custom house at Yorktown after that office had been established.

Because of a general complaint by masters of ships that there

36

were neither pilots nor beacons to guide them in Virginia waters, the General Assembly appointed Captain William Oewin chief pilot of James river in March, 1661, to be paid five pounds sterling for the pilotage of all ships above eighty tons if he be employed, and if not employed due to the presence of the ship's pilot who guided the vessel, he received forty shillings. The pilot was required to maintain good and sufficient beacons at all necessary places, and toward this expense, the master of every vessel that anchored within Point Comfort, having or not having a pilot, was required to pay thirty shillings. Later the pilot or the company to which he belonged was required to keep one pilot boat of 18 foot keel at least, rigged and provided for use at all times.

Early Ferries in Virginia

During the first quarter of the seventeenth century, the settler in Virginia used any kind of craft he possessed to cross the streams that separated him from his neighbor or for transacting business. Canoes, flatboats, scows, even sailing boats were pressed into service. These he propelled himself until he acquired a slave or two. Communication was aided by bridges across the smaller streams, and when horses became available, by crossing the rivers at the fords whenever possible.

The steady increase of settlers, however, created a demand for public transportation across creeks and rivers at the most travelled points. One of the first public ferries on record was started as a private enterprise in 1636, by Adam Thoroughgood. A skiff was rowed by slaves across the waters of Lower Norfolk, between what are now the cities of Norfolk and Portsmouth. In a few months the demand for transportation became so strong that the ferry was taken over by the county, increased to three hand-powered vessels and supported by a levy of six pounds of tobacco on each taxable person in the county.

A second early ferry was that of Henry Hawley in 1640, when

he was granted a patent by the court to keep a ferry at the mouth of the Southampton River in Kequoton, now Hampton, for the use of the inhabitants and other passengers during his natural life, not exacting above one penny for ferriage according to the offer in his petition.

"For the more ease of travellers," it was enacted by the General Assembly in January 1642, that the country provide and maintain ferries and bridges and the levy for payment to the ferrymen be made by the commissioners where the ferry is kept. This Act, establishing ferries at public expense, was repealed later and the court of each county given power to establish a ferry, or ferries in the county where needed at the instance of individuals. The court had authority to appoint and license the ferry keeper, to require of him a bond of twenty pounds sterling payable to His Majesty as security for the constant use and well-keeping of the boats. It was the duty of the court to order and direct the boats and hands in use at the ferries.

To encourage men to engage in operating ferries, it was enacted in 1702 that all persons attending on ferryboats should be free from public and county levies and from such public services as musters, constables, clearing highways, impressment, etc., and should have their licenses without fee or paying a reward for obtaining them. And if the ferryman desired to maintain an ordinary (public inn) at the ferry, he should be permitted to do so without fee for the license, but should be required to give bond for security. No other person should be permitted to establish an ordinary within five miles of such a ferry keeper. A warning was issued that any person not a ferryman who for reward should set any person over the river where there was a ferry, except for going to church, should pay for every such offense five pounds sterling, one-half to go to the ferryman and one-half to the informer, the full amount to the ferryman should he be the informer.

The county court was authorized in 1705 to make an agreement with the keeper of the ferry to set over the county militia on muster days and to raise an allowance for this in the county levy. All public messages and expresses to the government were to be allowed to cross ferry free. The adjutant general with one servant and their horses were exempted in 1738 from any payment on any ferry in the colony. Ministers of the church were likewise exempt from paying ferriage.

Dugout canoes of the Indians were among the first ferries used in Virginia and when more space was needed, two canoes were lashed together and secured by means of heavy cross pieces. In the *Journal* of Thomas Chalkley, a traveller in Virginia, he tells of a ferry crossing made at Yorktown in 1703: "We put our horses into two canoes tied together, and our horses stood with their fore feet in one and their hind feet in the other." Later, flatboats, scows, barges, and more carefully planked boats were put into use. Rope ferries were necessary wherever the current was swift, but used as little as possible on navigable rivers because of the obstruction to navigation.

The number of ferries in the colony increased steadily from year to year. At nearly every session of the General Assembly some law was enacted "for the good regulation of ferries." In 1705, the Assembly published a list of ferries with corresponding rates of ferriage that crossed the James, York, and Rappahannock Rivers and their branches. The ferries but not the rates are given herewith as follows:

Ferries on JAMES RIVER and branches thereof—
Henrico county at Varina.
Bermuda hundred to City Point.
Charles City county at Westover.
Appomatox river near Col. Byrd's store.
Prince George County at Coggan's point, and Maycocks.
Powhatan town to the Swineherd landing.
Surry county, Hog island to Archer's Hope.
Sicamore landing by Windmill point to the widow Jones's landing
  at Wyanoke.

Mouth of the Upper Chipoake's creek over to the Row, or Martin's
  Brandon.
Swan's point to James Town.
Crouche's creek to James Town.
James City county at James Town to Swan's point.
James Town to Crouche's creek.
Williamsburg, Princess Ann port to Hog island.
Chickahominy, at usual place on each side of river.
John Goddale's to Williams's neck, or Drummond's neck.
Nansemond county, Coiefield's point to Robert Peale's near Sleepy
  hole.
Elizabeth City county at Hampton Town from Town point to
  Brookes's point.
Hampton Town to Sewell's point.
Norfolk town to Sawyer's point or Lovet's plantation.

Ferries on YORK RIVER and branches—

New Kent county, Robert Peaseley's to Philip Williams's.
Brick House to West point.
Brick House to Graves's.
King William county, Spencer's over to the usual landing place.
Thomas Cranshaw to the usual landing place.
Philip Williams's to Peaseley's point.
West point to Brick House.
Abbot's landing over Mattapony river.
West Point to Graves's.
York Town to Tindal's point (Gloucester Point). This ferry was in
  continual operation until 1952 when a fine new bridge was opened
  for travel across the York. The ferriage in 1705 was seven pence
  half penny for a man, fifteen pence for man and horse.
Queen Mary's port at Williamsburg to Claybank creek in Gloucester
  county.
Captain Matthews's to Capahosack.
Tindal's point to York town.
Capahosack to Matthews's landing or Scimmino creek.
Bailey's over the Peankatank.
King and Queen county, Graves's to West point.
Graves's to Brick house.
Burford's to old Talbot's.

Captain Walker's mill landing.
Middlesex county, over Peankatank at Turk's ferry.

Ferries on the RAPPAHANNOCK RIVER—

Middlesex county, Shelton's to Mottrom Wright's.
Brandon to Chowning's point.
Essex county, Daniel Henry's to William Pannell's.
Bowler's, at the usual place, to Sucket's point.
Tappahannock to Webley Pavies, or to Rappahannock creek.
Henry Long's to the usual place.
Richmond county, William Pannel's over the Rappahannock.
Sucket's point to Bowler's.

<h2 align="center">POTOMAC RIVER—</h2>

Stafford County, Col William Fitzhugh's landing to **Maryland**.

<h2 align="center">EASTERN SHORE—</h2>

Port of Northampton to the port of York.
Port of Northampton to the port of Hampton.

Rates on these ferries were fixed by courts and varied according to distance. Across the Southampton River in Hampton the rate was one penny, while from the Port of Northampton to Hampton, the price was fifteen shillings for a man and thirty shillings for a man and horse.

In 1740, the ferry from Hampton to Norfolk was described as follows: "From the town of Southampton, across the mouth of the James River, to the borough of Norfolk and Nansemond town; from the borough of Norfolk and Nansemond town, across the mouth of the James river, to the town of Southampton." The fare for this trip for a man passing singly was seven shillings, six pence; for a man and horse, five shillings each.

By February 1743, the ferries across the Chesapeake Bay had been expanded, and were described as follows: "From York, Hampton and Norfolk towns, across the Bay to the land of Littleton Eyre on Hungar's river in Northampton County; from the land of Littleton Eyre on Hungar's river in Northampton

41

County, across the Bay to York, Hampton and Norfolk." The rate for a man was twenty shillings, for a man and horse, fifteen shillings each.

In 1748, another list of ferries, published in Hening's *Statutes*, showed that the number had more than doubled since 1705. The Potomac river had added fourteen to the number given at that time. Two ferries had been established on Nottaway: "From Thomas Drew's land to Dr. Brown's, and from Bolton's ferry to Simmons' land." The ferries in addition to those of 1705 are the following:

JAMES RIVER and branches—

Land of Henry Batte in Henrico County, to the Glebe land at Varina.

Westover in Charles City county, to Maycox, or Coggins point, and from Maycox to Westover.

Kennon's to Maye's on Appomattox river, and from Maye's to Kennon's.

Joseph Wilkin's or John Hood's land in Prince George county, to John Minge's land in Wyanoke.

Hog-Island, in Surry county, to Higginson's landing on Col. Lewis Burwell's land.

Jamestown to Swan's Point.

Cowle's to Williams's.

Cowle's to Hamner's point.

Crawford's to Powder point.

Bolling's point in Henrico county, over Appomattox river.

City point to Shirley hundred, at the ship landing, and from the said landing to City Point.

Ship landing at Shirley to Bermuda hundred.

Bermuda hundred to City Point.

Hemp landing at the falls of James river, to Shocoe's, on the land of William Byrd, esq.

Land of Stephen Woodson, in the county of Goochland, to Manacon town.

Henry Cary's land, over the river, to the land of the said Cary.

Henry Batte's, in the county of Henrico, to Alexander Bollings, in the county of Prince George.

Land of Col. Richard Bland, in the county of Prince George, to the land of Mrs. Anderson, in the county of Charles City.

Land of William Pride called the store landing, in the county of
  Henrico, to Anthony's landing, in the county of Prince George.
Store landing over Persie's stile creek, to the land of Peter Baugh.
Warehouse landing at Warwick, to the land of Thomas Moseley.
Mulberry island point in the county of Warwick, to Cocket's in Isle
  of Wight, and from Cocket's to Mulberry island.
Land of Richard Mosby in Goochland county, to the land of Tarlton
  Fleming, opposite to Mosby's landing.
Land of Tucker Woodson, to the land of Paul Micheaux near the
  court house.
Land of Bennet Goode to the land of Col. John Fleming.
Land of James Fenly to the land of William Cabell, cross the Flu-
  vanna.
Charles Lynch's plantation in Albemarle county, on the Rivanna,
  cross the said river, to the land of Richard Meriwether.
Land of Mr. Benjamin Cocke, cross the said river, to the land of the
  said Benjamin Cocke.
Land of Ashford Hughes on the north side of James River, near the
  mouth of Willis Creek, cross the river to the land of Robert Carter,
  and from the said Carter's to the said Hughes's.
Land of Lemuel Riddick, adjoining the public wharf in Suffolk, cross
  Nansemond river, to Samuel Jordan's land.
Land of William Pride in the county of Herrico, on Appomattox
  river, above the narrow falls, to the land of the said Pride over the
  river, in Prince George county.
Land of William Cabbell, in Albemarle county, at the mouth of
  Swan's creek, over the Fluvanna, to the land of Samuel Spencer;
  or from the said Cabbell's, over Tye river, to his land opposite.

Additional ferries on the YORK RIVER—

Chamberlayne's to Williams's.
Brick House to Dudley's, or Dudley's to Brick House.
Webb's to Lyde's, formerly Spencer's, in King William county.
Temple landing, over Matapony river.
West Point to Dudley's, or Dudley's to West point.
Capahosic to Scimino.
Seaton's over Pianketank.
Frazier's to Broach's, and from Broach's to Frazier's.
Walker town to Waller's, or Waller's to Walker town.
Turk's ferry over Pianketank.

Robert King's over Pamunky to Blackwell's, or from Blackwell's to
    King's.
Sweethall to Claiborne Gooch's, or from Claiborne Gooch's to Sweet-
    hall.
George Dabney's over Pamunky river.
Taylor's in King William to Garland's in Hanover.
William Pulliam's in Hanover, to John Holliday's in Caroline.
Richard Littlepage's to Thomas Claiborne's land, over Pamunky, and
    from Claiborne's to Littlepage's.
Todd's warehouse landing, in King and Queen, to the land of Robert
    Armistead Bird, in King William.

Ferries on the RAPPAHANNOCK RIVER—

Whiting's to Gilbert's.
Land of Thomas Ley to Robinson's, or from Robinson's to Ley's.
Byrd's to Williams', or Williams' to Byrd's.
Tappahannock town to Carter's, or to Rappahannock creek, on either
    side thereof.
Tankersley's over Rappahannock river, to the usual place.
Germanna over the Rapid Ann.
Ray's plantation to Skinker's.
Urbanna to Chetwood's.
Urbanna, from the ferry landing to Locust point, on the land of
    Ralph Wormley, esq.
Johnston's plantation in Spotsylvania, to Washington's in King
    George.
Taliaferro's plantation of the Mount, to the land of Joseph Berry.
Philemon Cavenaugh's ford.
Wharf above the mouth of Massaponax creek, to the opposite land-
    ing upon Mr. Ball's land.
Fredericksburg warehouse to the land of Anthony Strother, or from
    Strother's to Fredericksburg.
Roy's warehouse to Gibson's warehouse.
William Lowry's to the land of Benjamin Rust, or from Rust's to
    Lowry's.
Falmouth to the land of Francis Thornton, in Spotsylvania.
Hackley's land in King George to Corbin's in Caroline.
Lot of Joseph Morton, in Leeds town, to the lands of Mrs. Brooke.
Lower side of Parrot's creek to Teague's creek, on the land of Baldwin

Matthews Smith, and from that creek to the lower side of Parrot's creek.

Ferries on the POTOMAC RIVER—

Col. William Fitzhugh's land at Boyd's hole, over to Maryland.

Hoe's to Cedar point.

Tripplet's land below the mouth of Quantico creek, over to Brooks's land.

Robert Lovell's in the county of Westmoreland, over to Maryland.

Land of William Russel on Sherendo, cross into the fork, or cross the main river.

Kersey's landing on Col. Carter Burwell's land, to the land of Col. Landon Carter.

Gersham Key's land, to the land of the Honourable William Fairfax.

Williams' Gap, from the land of the Right Honourable the Lord Fairfax, where John Melton now lives, to the land of Ralph Wormley, Esquire.

Plantation of George Mason, opposite to Rock creek, over to Maryland.

Plantation of John Hereford in [Doegs?] neck, over the river, to the lower side of Pamunky in Maryland.

Hunting creek warehouse to Frazier's point, or Addison's.

Land of Ebenezer Floyd to Powell's.

Evan Watkin's landing, opposite to Canagochego creek, to Edmund Wade's land in Maryland.

Land of William Clifton to the land of Thomas Wallis.

Land of Hugh West to Frazier's, or Addison's.

The county courts were required to appoint proper boats to be kept at the ferries where needed for the transportation of wheeled vehicles—carts, chaises, coaches and wagons. The rates for these vehicles were based upon the rates for horses. For every coach, chariot or wagon, the price was the same as for the ferriage of six horses; for every cart or four-wheeled chaise, the price was the same as for four horses; and for every two-wheeled chaise or chair, the same as for two horses. For every hogshead of tobacco, the rate of one horse was charged. For ferrying animals, every head of neat cattle rated as one horse; every sheep, lamb or goat,

45

one-fifth part of the rate for a horse; for every hog, one-fourth of the ferriage of a horse.

Should the ferryman exceed the legal rates, he was penalized by having to pay to the party aggrieved, the ferriage demanded and ten shillings. In February 1752, a free ferry for any persons and their commodities was established from the town of Port Royal over the Rappahannock river to the land of John Moore in King George County. In 1757, there were five ferries from Norfolk over her various bodies of water, one of which was established as a free ferry supported by the county to enable the poor people of the community to have free passage to market.

In the *Virginia Gazette* for March 31, 1768, the following advertisement appeared: "I have boats for the use of my ferry equal to any in the government, and can give ferry dispatch greater than any other ferry keeper on the Potomac river." In the late seventeenth century, the Henrico county ferry was run by a woman. The county levy for that year was the sum of 2,000 pounds of tobacco to be paid to Mrs. Sarah Woodson for keeping the ferry for one year.

The county courts continued to establish new ferries and to discontinue others through the Revolution and after. Now and then bridges would take the place of ferries across the smaller streams. An interesting instance of such a change is told in the *Richmond Times-Dispatch* for August 20, 1939. "For a century from 1650, ferries were maintained across the two branches of Pagan river at Smithfield in Isle of Wight county. In 1750, these ferries were abandoned for toll bridges." From year to year, ferries gradually gave way to bridges and now, when we have passed the middle of the twentieth century, there are few ferries left in Virginia. These are large, fine steamboats capable of carrying hundreds of passengers, but are no more necessary to the welfare of the people than were the little dugouts in the early days of the colony.

Shipbuilding in the Period of the Revolution

At a Convention of delegates and representatives of the counties and corporations of the Colony of Virginia on July 17, 1775, there was established a Committee of Safety consisting of ten prominent men for putting into execution the ordinances and resolutions of the Convention. That committee was authorized to provide as many armed vessels as they judged necessary for the protection of the Colony in the war that seemed to threaten. Advertisements for ship-carpenters and other operatives were made, and every inducement held out to them in order that the building of vessels might immediately commence.

Between December, 1775, and July, 1776, the Committee established a small navy by purchase of several armed, schooner-rigged vessels from the owners of the merchant fleet; and contracts were made for a number of galleys to be constructed on the different rivers of the Colony. The Potomac was to be protected by the construction of two row-galleys and the purchase of three boats. George Minter was elected master of a row-galley to be built on the James River under the direction of Colonel Cary. He was requested to recommend proper persons to be mate, two midshipmen, gunner, and to enlist forty seamen.

John Herbert, a master shipbuilder, was employed to engage any number of ship-carpenters that he could procure upon reasonable terms, and to examine such places upon the James River or its branches as he thought proper and convenient for erecting shipyards, and to report to the Committee.

Caleb Herbert was retained as the master builder of a shipyard on the Rappahannock River, and Reuben Herbert for such a yard on York River. Each of them was desired as soon as possible to engage a proper number of workmen for building two row-galleys to be employed in the two rivers to transport troops. It was recommended that a committee at Norfolk engage a proper person to take direction and employ a number of ship-carpenters for at least a year, to build vessels for the Colony.

George Mason, in a letter to George Washington on April 12, 1776, mentioned that he had under his charge two row-galleys of 40 or 50 tons burden, each to mount light guns, three and four pounders; and the sloop, *American Congress*, a fine stout vessel of 110 tons burden, mounting fourteen carriage guns, four and six pounders, and was considering mounting two 9-pounders upon her main boom.

On June 6, 1776, the Committee of Safety appointed Christopher Calvert to superintend the building of two row-galleys for the protection of Virginia and North Carolina, to engage a master workman and as many men as he should need to work expeditiously. The two vessels, *Caswell* and *Washington*, were built at the South Quay Shipyard on the Blackwater River near the North Carolina line. A North Carolina sloop had been seized in Ocracoke Inlet in April, 1776. Sometime later, a warrant for £100 was issued to Argyle Herbert for the use of Captain Calvert upon account to pay the carpenters employed on his galley.

At the convention of delegates held at the Capitol in Williamsburg on May 6, 1776, resolutions were passed dissolving the Government from Great Britain, establishing Virginia as a Commonwealth or State. A Board of Navy Commissioners composed of five members was appointed to superintend and direct all matters relating to the Navy. Their peculiar duties were defined as follows: To superintend and direct the building and repairing of all vessels; provide the necessary outfits, ordnance, provisions and naval stores; control the public rope walks; erect dockyards; contract for and provide all timber necessary for building purposes; and supervise the shipyards.

On September 12, 1776, this Commission was requested to engage the proper persons for building "in the most expeditious manner", 30 boats for the transportation of troops on the rivers, each boat to be the proper size for carrying a complete company of 68 men with their arms and baggage. Those were small boats without masts but broad and strong enough to transport troops across rivers and to carry from point to point large quantities of ammunition and provisions as they were required. The small boats had been found indispensable in retreats, in rapid marches, and in concentrating land forces.

The Commissioners were authorized in October to provide the necessary plank and timber for the building of four large galleys fit for river and sea service, and to be mounted with proper guns. And for manning these galleys and others being built, the Commissioners were requested to raise the number of men needed, not to exceed 1300 to serve three years.

The Continental Congress directed that two frigates of 36 guns and of 500 tons burthen be built in Virginia, and the Navy Board ordered the work done at Gosport Shipyard in Norfolk County. The following excerpts from a letter of Richard Henry Lee of the United States Congress to James Maxwell, Chief Superintendent of Construction on December 1, 1776, give directions for building the frigates:

The Congress has resolved upon building two ships-of-war of 36 guns each. . . . You, Sir, have been recommended as a person of great fitness for this business. . . . I do, in the name of the committee, request you will . . . determine a most fit place to put these ships upon the stocks at. Safety against the enemy is a very necessary object, proper water for launching, and convenience for getting timber you will consider. . . . A master builder with four or six workmen will soon go hence to Virginia for this business, and I have no doubt other workmen will be had in that State to carry on the work briskly. . . . The builder desires that trees be felled immediately whilst the sap is down, that a quantity of locust trunnels be split one and one-half inches and from 18 to 30 inches in length; that sawyers be employed to get out white oak plank of 3½ inches. These things and whatever else may be immediately necessary for this business you will take care to have done. . . . The builder tells me that cedar, locust, pitch pine, or wild cherry will be the proper timber for the upper works.

On Wednesday, December 18, 1776, it was resolved by the General Assembly that the Governor be desired to write to the Maryland Council of Safety to inform them that four galleys of eighty odd feet keel, intended for the protection of Chesapeake Bay and adjacent capes and coasts, were then building in Virginia and in great forwardness, and that the General Assembly have directed four more galleys, much larger, be immediately built and equipped for the same purpose. The hope was expressed that the sister state, equally interested in mutual defence, would supply a proper quota of galleys to act in concert with those of Virginia. Chesapeake Bay was the chief theatre of action by the enemy because of the principal tories residing near its waters. To watch their movements and prevent intercourse with the enemy became the duty of these galleys.

Two galleys, the *Accomac* and *Diligence,* were built in 1777 on Muddy Creek near Guilford in Accomack County, and stationed on the Eastern Shore. These large galleys were about 90 feet in length and each carried two 18-pounders, four 9-pounders, and several swivels, in all ten guns.

The State built and operated in 1777, a ropewalk at Warwick in Chesterfield County about five miles below Richmond, where ducking, sail-cloth, and rope were manufactured under the charge of Captain Charles Thomas. Several important warehouses had been established there. The place was totally destroyed in the British raid of April, 1781.

There were numerous places in Virginia where shipbuilding was carried on during 1776 and 1779. Vessels were built and equipped on the Eastern Shore, the Potomac, the Rappahannock, Chickahominy and James Rivers; at Hampton, Gosport in Norfolk County, South Quay on the Blackwater near the Carolina line, Frazier's Ferry on the Mattaponi, and Cumberland on the Pamunkey. This last shipyard was discontinued at the suggestion of Thomas Jefferson in 1779 because of the enormous expense attending its support. There was also a shipyard in Gloucester

County owned by John Hudgens. Construction was carried on chiefly at the Chickahominy and Gosport yards.

The shipyard on the Chickahominy was located about twelve miles from its mouth and chosen partly because of its sheltered location and the fine timber that grew near by. The Navy Board had purchased 119 acres of land for the sum of £595 in April, 1777, and it became one of the busiest shipyards in the State. The ship *Thetis,* and the armed brig *Jefferson,* and many others were built in this yard. This establishment suffered the same fate as the Warwick ropewalk during Arnold's raid in 1781. A few posts are still standing in the water to mark the spot.

Just before the breaking out of the Revolution, the British Government had established a marine yard at Portsmouth, Virginia, for the use of its Navy, and named it for the dockyard Gosport near Portsmouth, England. This yard was confiscated by Virginia when the war began, and enlarged in 1801, by the purchase of 16 acres of the estate of Andrew Sproule, the British Navy Agent, for $12,000. The ship *Virginia* was built here and the two frigates laid on the stocks, with a number of other vessels.

Early in May, 1779, a British fleet with a large force of frigates and transports passed through the Capes and on into Hampton Roads, under the command of Sir George Collier. Unable to meet such a formidable enemy, the Virginians withdrew their small fleet up the river for safety. The following extract is said to be from the *Journal* of H.M.S. *Rainbow,* commanded by Sir George Collier:

When the troops under General Matthews took possession of Portsmouth, Norfolk and Gosport Navy Yard had been abandoned. Before leaving, the Virginians had set fire to a ship-of-war of 28 guns ready for launching, belonging to Congress, and two French merchant ships loaded with bales of goods and tobacco. . . . The quantities of naval stores found in their arsenals were astonishing. Many vessels of war were on the stocks in different stages of forwardness; one of 36 guns, one of 18, three of 16, and three of 14, beside many merchantmen. The whole number taken, burnt, and destroyed while the King's ships were in the river amounted to *one hundred and thirty-seven* sail of vessels. . . . [Evidently, James Maxwell's two frigates were included in this group.] Five thousand loads of fine seasoned oak knees for shipbuilding and an infinite quantity of plank, masts, cordage, and numbers of beautiful ships-of-war on the stocks were at one time in a blaze and totally consumed, not a vestige remaining but the iron work. . . . Quantities of tar were found in the warehouses, and in Suffolk, 8,000 barrels of pitch, tar, and turpentine were seized. Much was carried away but great quantities were set on fire and left behind.

Early in 1780, it was learned that the enemy intended another invasion of the coast of Virginia, and the General Assembly took measures for defense. In addition to land forces, the Navy was ordered to assemble a small fleet consisting of the ships *Thetis, Tempest,* and *Dragon,* the brig *Jefferson* and the galley *Henry* for the purpose of defending Hampton Roads and adjacent waters. In October, the situation seemed much more critical and Acts were passed to build two more galleys of the same construction as built by Congress in 1776, carrying two 32-pounders in the bow, a like number in the stern, with 6-pounders at the sides. The rigging, sails, guns, and other materials to be provided while the galleys were on the stocks that no time be lost in preparing them for the cruise.

Captain James Maxwell addressed a letter to Governor Jefferson on December 7, 1780, informing him that the Lieutenant of the *Jefferson* thinks it will take £14,000 [in continental money] to pay her up to the present time. There was also due the workmen of the Gosport Shipyard on the last of October, £18,679-14s-6d. Clothing was wanting for 26 men—52 shirts, 26 jackets, and breeches, stockings, shoes and hats or caps.

Governor Jefferson wrote to James Maxwell on January 16, 1781, as follows: "I enclose you a plan for building portable boats, recommended by General Washington, and shall be glad that you will take measures for having about twenty of them made without delay. We have doubts that they will suit our waters, and will be glad to confer with you on any suggested improvement."

General Lafayette having arrived at York on March 13, 1781, Governor Jefferson wrote him that there would be ready for him at the Chickahominy Shipyard four boats well-fitted to his purpose, and others were collecting in the rivers to rendezvous at Hood's. These were for lookout boats placed in the Rappahannock, Piankatank, and York Rivers. Hood's was a battery on the James in Prince George County, opposite Weyanoke, now called Fort Powhatan. Later, Maxwell notified the Governor that he was building a few boats at the Chickahominy Shipyard. The Governor had requested that a good bateau builder be sent there to superintend some carpenters in building bateaux for the river above the Falls, and the rest of the carpenters be set to building boats for navigating the lower parts of the river, boats so light and of such form they could be moved on wheels.

On April 21, 1781, the traitor Arnold and Phillips made their raid up the James River, penetrating as far as Richmond. A detachment under Lieut. Col. Ambercrombie destroyed the shipyard at Chickahominy including a large number of naval craft, among them an unfinished ship of 200 tons, and important warehouses. On April 27, the Virginia fleet composed of six ships, eight brigs, five sloops, two schooners and several smaller craft, met the British fleet in battle a few miles below Richmond, but had to give way. A number of vessels were scuttled or set on fire, but the enemy captured the rest, and the fleet was practically wiped out. Only one armed vessel remained, the brig *Liberty*.

After the surrender of Cornwallis, the General Assembly met in May, 1782, and appointed three Commissioners to superintend the work of protecting the Bay. The ship *Cormorant* and the brig *Liberty* were prepared, and plans made for building two galleys and two barges or whale boats. The Commissioners managed to keep a small naval force together during 1782 and 1783, until the war came to an end. When peace was declared in 1783, the Commissioners had in different stages of construction the schooners *Harrison* and *Patriot,* the barges *York* and *Richmond,* and the pilot boat *Fly.* Virginia dispensed with all her fleet except the *Liberty* and *Patriot* which were retained, with the approval of Congress, as revenue cutters.

Among the various types of vessels mentioned here, galleys are generally thought of as having been rather insignificant. On the contrary, they were among the important vessels constructed for the Virginia Navy. While they were so built that they could easily retire up the creeks out of range of British guns, they were capable also of sailing out in the broad waters of the Bay. They were broad in proportion to their length which varied from 60 to 90 feet, and not drawing much water could support immense weight upon their decks, as in transporting troops with their horses and baggage, and in carrying guns of the largest size. Generally they had two masts and were rigged as schooners, but an occasional galley carried three masts as in the case of the *Gloucester.* Some were without masts and were called row-galleys. These were only half decked, were provided with high and strong bulwarks for the better protection from marksmen, and were propelled by oars only.

The armaments of these galleys were much more formidable in proportion to their tonnage than were those of any other vessels. In November, 1776, two large galleys for river and sea service were ordered to be built to carry four 24-pounders, and fourteen 9-pounders each. Also, in October, 1780, two more large ones were ordered to carry two 30-pounders in the bow, the same in the stern, with 6-pounders at the sides, for the protection of the Chesapeake Bay.

The *Gloucester* was one of the largest galleys built. Judging from the order sent to Captain Charles Thomas on April 30, 1777, for rope and cables from the ropewalk at Warwick, the galley had a foremast, a mainmast, a mizzen and a bowsprit. All the rigging was to have a rogue's yarn in it, that it might be distinguished from merchant rope. A rogue's yarn was a single thread of red or blue which was twisted in the rope at the manufactory, and served to distinguish it from all others. The *Gloucester* was used as a prison ship.

Two accounts of the development of the schooner in use by Virginia during the Revolution are worth recording:

(a) It is from this time perhaps that we may date that new era in the art of shipbuilding which now produced the firstlings of that brood of fast-sailing clippers that afterwards were to astonish and charm the naval world with their brilliant performance. The Americans were the originators of this improved naval architecture. It was developed by that spirit of invention and love of adventure so characteristic of a young and vigorous people, urged by necessity. . . . The far-famed Baltimore clipper soon established the reputation of that long, low, rakish-looking craft, which has ever since been the cynosure of the seaman's eye.

(b) The most spectacular event in the history of naval architecture in the 18th century was the emergence of the clipper-schooner which became famous during the Revolution. This was a trim, rakish craft known as the Virginia-built schooner, an exclusively Chesapeake type prior to the Revolution. The war created a demand for this fast-sailing vessel and builders all along the coast constructed vessels on the clipper lines thereby converting it to a national type. The war made the clipper-schooner internationally known, however, and before the end of the century, the French, Dutch, and British built schooners on the clipper lines.

The pilot boat used in the Virginia Navy was a small fast-sailing craft used as "lookouts", only two of which, the *Molly* and the *Fly,* were armed. Their duties were attended with many hardships and extreme peril. They were obliged to hover along a dangerous coast in all weathers to give notice of the approach of every sail whether friend or foe. They acted as a flying sentry at the gates of the Chesapeake, but constantly exposed to the broad Atlantic outside.

Although the war virtually eliminated Virginia's trading fleet
as well as her Navy, her shipbuilding capacity was at its best.
Her many shipyards, abundant supplies of available shipbuilding
timber, and her skilled craftsmen soon put her trading fleet in
operation and it became an integral part of the American Merchant Marine.

EARLY VIRGINIA WATERCRAFT
(as defined by authorities)

Shallop—A nondescript type of small boat, from the French "chaloupe," open or half-decked, sometimes with one or two masts for use if needed. It was the most popular boat used in the colony for collecting corn from the Indians, fishing, oystering, and exploring.

Pinnace—"An old name in English marine nomenclature." A light sailing vessel from the sixteenth to the eighteenth century, decked and having one or more masts, from twenty to thirty tons burden. The pinnaces *Virginia, Discovery,* and the two built at Bermuda, *Deliverance* and *Patience* were sea-going vessels.

Barge—"A term applied to numerous types of vessels throughout the ages." In Virginia it meant a ship's boat, or a flat bottom freight boat used on inland waterways and for loading and unloading ships.

Bateau—The Chesapeake Bay bateau in colonial times was a double-ended boat having a V-bottomed hull, built in lengths to forty or fifty feet, and was primarily a rowing or poling boat used for rivers and creeks.

Scow—A large flat-bottomed vessel having broad, square ends and straight sides, sometimes flat-decked. Probably from the Dutch term "schouw."

Flat—An old form of boat, simple to build, with flat bottom, ends boarded over, used for heavy freight and ferrying, sometimes having a mast.

Skiff—A light swift open boat, generally double-ended for rowing, but sometimes equipped for sailing.

Frigate—Originally a light vessel propelled by both sails and oars with flush decks. A "frigott" was constructed at Cape Comfort by Captain Argall in 1613. Later the term was applied only to a type of warship.

Punt—A small flat-bottomed, open boat, usually with a seat in the middle, and a well or seat at one, or each end for use in shallow waters, propelled by oars or poles.

Yawl—A small sailing vessel rigged like a sloop with a small additional mast in the stern.

Canoe—The evolution of the Chesapeake Bay canoe and the Chesapeake Bay bugeye from the Indian dugout canoe, is one of the most interesting developments in the history of shipbuilding in America.

Piragua or *Periagua*—A large dugout canoe fitted with sails.

Tobacco Boat—The double dugout canoe generally referred to as the tobacco boat, was "invented" by the Reverend Robert Rose, rector of St. Ann's Parish in Albemarle. The boats were from fifty to sixty feet in length, from four to five feet in width, clamped together with cross beams and pins, two pieces running lengthwise over these, with a capacity of from five to ten hogsheads of tobacco. The first mention of this boat was in Rose's diary for March 14, 1749. (2) The James River bateau or tobacco boat was invented by Anthony J. Rucker in 1771, and is mentioned in Jefferson's *Notes on Virginia*. The bateaux were made of boards from forty to sixty feet long and flat-bottomed. They were constructed so that either end could be poled against the river bank and the hogshead rolled aboard. Each craft required a crew of three, one to steer and one each for the sideboards, the full length of the gunwales.

Sloop—A craft with a single mast and fore-and-aft rig, in its simplest form a mainsail and jib. It is said to have appeared in the colony from England before 1630, and became the most common colonial rig. It was the fast-sailing craft for coastwise and West Indies trade. It became very popular as a pleasure boat.

58

Schooner—A two or more masted vessel, fore-and-aft rigged. The essentials of the schooner are two fore-and-aft sails and a headsail (jib), any other sails being incidental. This type of rig was not known until the last quarter of the seventeenth century, appearing in America by 1700, or shortly after. During the second half of the eighteenth century, the schooner displaced the sloop as the principal colonial coasting vessel, and during the Revolution emerged as the most distinctly American type.

Pilot Boat—In 1661, the General Assembly passed an Act creating the office of Chief Pilot of the James River. A specific type of vessel evolved for use as pilot boats—fast, weatherly boats, somewhat on the mold of the already developing clipper schooner, about 1745. This boat soon acquired schooner rig and all the characteristics of a clipper schooner. This trim craft, distinguished for speed and sea worthiness, proved ideal for yachting. Almost all schooner yachts until about 1870, were built on the lines of pilot boats. The best known example was the victory of the yacht *America* in 1851.

Brig—A seagoing vessel having two masts and square rigged.

Brigantine—A seagoing vessel having two masts, one square rigged, the other fore-and-aft.

Snow—A seagoing vessel having two masts similar to a brig, and an additional mast abaft the mainmast which carried a spanker or driver (a gaff-headed trysail).

Ship—A sailing vessel having three or more masts, square rigged, the largest seagoing vessel of the period. A term frequently applied to any vessel.

Bark or *Barque*—A sailing vessel having three or more masts, square rigged, the after mast, fore-and-aft rigged. A term frequently applied to any vessel.

Barkentine—A sailing vessel with three or more masts, the fore mast square rigged, the other masts being fore-and-aft.

Galley—A long, single or partially decked vessel of light draft, fitted for rowing and having one or two masts to raise for use when needed. They ranged in size from forty to seventy-five feet in length, and were used as warships by Virginia during the Revolution when they carried from one to twelve guns.

The planters and shipbuilders of Virginia had a wide choice in the selection of timber for building their boats and ships:

Virginia yielding to no known place in the known world for timbers of all sorts, commodious for strength, pleasant for sweetness, specious for colors, spacious for largeness, useful for land and sea, for housing and shipping. For timber, we have the oak, ash, poplar, black walnut, pines and gum trees.

Frequently several kinds of wood were used in the construction of a boat, and the color combinations of the natural woods, with the use of turpentine and pitch, was pleasing enough to some shipbuilders. For others, however, the vessels were painted in bright colors, often a combination of several colors. The larger vessels were usually built of white oak, but due to the rapid growth of the tree, Virginia oak was not as good or lasting as the oak grown in England. Ships built from the American live oak, helped much to improve the reputation of colonial vessels.

As a general rule, vessels built in the colony were without ornamentation of any kind, utility being the watchword, and speed important. It has been reported, however, that a few billet heads and figureheads were placed on ships, and carved figureheads imported from Boston by a planter appeared on his vessels.

BIBLIOGRAPHY

Abbot, W. J. *American Merchant Ships and Sailors*. New York, 1902.

Ames, S. M. *Studies of the Virginia Eastern Shore in the Seventeenth Century*. Richmond, 1940.

Andrews, C. M. *The Colonial Period of American History*. New Haven, Yale University Press, 1934-1938. 4 vols.

Beverley, Robert. *The History and Present State of Virginia*. London, 1705. Reprinted for *The Institute of Early American History and Culture* by the University of North Carolina, 1947.

Bishop, J. L. *The History of American Manufacturers from 1608 to 1860*. Philadelphia, 1861-1864. 2 vols.

Bloomster, E. L. *Sailing and Small Craft Down the Ages*. United States Naval Institute, 1940.

Bolton, H. E. *The Spanish Borderlands*. New Haven, 1921.

Brewington, M. V. *Baycraft Labels at Dorothy's Discovery*. Cambridge, Md., 1952.

Brown, Alexander. *The First Republic in America*. Boston, 1898.

———*The Genesis of the United States*. Boston, 1890. 2 vols.

Bruce, P. A. *The Economic History of Virginia in the Seventeenth Century*. New York, 1896. 2 vols.

Brumbaugh, G. M. *Revolutionary War Records. Vol. 1, Virginia*. Washington, 1936.

Byrd, William. *The Secret Diaries of William Byrd of Westover, 1709-1712, 1739-1741*. Richmond, 1942. 2 vols.

Calendar of State Papers, Colonial Series. 1574-1731 (not complete). London, 1860-1938. 18 vols.

Calendar of Virginia State Papers, compiled by William Price Palmer. Richmond, 1875-1893. 11 vols.

Campbell, Charles. *History of the Colony and Ancient Dominion of Virginia*. Philadelphia, Pa., 1860.

Chapelle, Howard I. *American Small Sailing Craft*. New York, 1951.

Chatterton, E. K. *English Seamen and the Colonization of America*. London, 1930.

Dictionary of American History. New York, 1940. 6 vols.

Fassett, J. F. G. *The Shipbuilding Business in the United States*. New York, *Society of Naval Architects and Marine Engineers*, 1948.

Fithian, P. V. *Journal and Letters*. Williamsburg, 1943.

Flippen, P. S. *The Royal Government in Virginia, 1624-1776*. New York, 1919.

Fry and Jefferson's *Map of Virginia*. 1775.

Grahame, James. *History of the United States of America from the Plantations of the British Colonies Until Their Assumption of National Independence*. Philadelphia, Pa., 1845. 4 vols.

Gwathmey, J. H. *Historical Register of Virginians in the Revolution. 1775-1783*. Richmond, 1938. "Vessels of the United States Navy," p. 861.

Hakluyt, Richard. *Principal Navigations, Voyages, and Discoveries of the English Nation*. Glasgow, 1905. 12 vols.

Hariot, Thomas. *A Brief and True Report of the New Found Land of Virginia*. Frankfort, De Bry, 1590. Also, a facsimile reprint of the first edition, 1588. New York, 1903.

Hening, W. W., ed. *Statutes at Large*. Richmond, Va., 1809-1823. 13 vols.

Herrera, Antonio de. *General History of the Vast Continent and Islands of America*. London, 1726. 6 vols.

Huntley, F. C. *The Seaborne Trade in Virginia in Mid-Eighteenth Century*. In *Virginia Magazine of History and Biography*, vol. 59.

Johnson, E. R., and Collaborators. *History of Domestic and Foreign Commerce of the United States*. Washington, 1915.

Johnson, Robert. *Nova Britannia*. 1609. (In *Force's Tracts*, vol. 1)

Kelly, Roy, and F. J. Allen. *The Shipbuilding Industry*. Boston, 1918.

Latane, J. H. *Early Relations of Maryland and Virginia*. Baltimore, 1895. (*Johns Hopkins University Studies*. 13th ser., III-IV).

Lefroy, J. H. *Memorials of Bermuda*. London, 1877. 2 vols.

Lull, E. P. *History of the United States Navy Yard at Gosport, Virginia*. Washington, 1851.

Mackintosh, J. *The Discovery of America and the Origin of the North American Indians*. Toronto, 1836.

Mason, F. N., ed. *John Norton and Sons, Merchants of London and Virginia*. Richmond, 1937.

Mason, P. C. *Records of Colonial Gloucester County*. Newport News, Va., 1946-1948. 2 vols.

Mereness, N. D., ed. *Travels in the American Colonies*. New York, 1916.

Middleton, A. P. *Tobacco Coast, a Maritime History of the Chesapeake Bay in the Colonial Era*, edited by George C. Mason. Newport News, 1953.

————New Light on the Evolution of the Chesapeake Clipper Schooner. (In *The American Neptune*, vol. 9)

Moore, G. M. *A Seaport in Virginia*. Richmond, 1949.

Morris, E. P. *The Fore and Aft Rig in America*. New Haven, Yale University Press, 1927.

Morriss, M. S. *The Colonial Trade of Maryland, 1689-1715*. (*Johns Hopkins University Studies in History and Political Science*. Series 32.)

Neill, E. D. *History of the Virginia Company*. Albany, N. Y., 1869.

Norfolk County Deed Book, vol. F. manuscript.

Palmer, W. P. *The Virginia Navy of the Revolution*. (In the *Southern Literary Messenger*, January-April, 1857.)

Paullin, C. O. *The Navy of the American Revolution*. Cleveland, 1906. "The Virginia Navy," pp. 396-417.

Percy, Alfred. *Piedmont Apocalypse*. Madison Heights, Va., 1949.

Purchas, Samuel. *Purchas His Pilgrimes*. Glasgow, 1905-1907. 20 vols.

Quinn, D. B. *The Roanoke Voyages, 1584-1590*. London, Hakluyt Society, 1952. 2 vols.

Ralamb, Ake Classon. *Skeps Byggerij Eller Adelig Ofnings*. Stockholm, 1691. Reprinted at Malmo, 1943.

Robinson, Conway. *Account of Discoveries in the West until 1519, and of Voyages to and along the Atlantic Coast of North America from 1520 to 1573*. Richmond, 1849.

Smith, John. *Works, 1608-1631*. Arber edition. Birmingham, 1884.

————Same, with introduction by A. G. Bradley. Edinburgh, 1910. 2 vols.

Spotswood, Alexander. *Official Letters*. Richmond, 1882. 2 vols.

Stewart, R. A. *The History of Virginia's Navy of the Revolution*. Richmond, Va., 1933.

Swem, E. G. *Virginia Historical Index*. Roanoke, Va., 1934-1936, 2 vols.

The Trades Increase. London, 1615.

Tyler, L. G. *The Cradle of the Republic*. Richmond, 1906.

Virginia (Colony). *Minutes of the Council and General Court of Colonial Virginia, 1622-1632*. Richmond, 1924.

Virginia Gazette. Williamsburg, 1736-1780.

Virginia Company of London. Records, edited by S. M. Kingsbury. Washington, 1906-1935. 4 vols.

Virginia. Governor. *Official Letters of the Governors of the State of Virginia*. Vol. I: Patrick Henry, July, 1776-June, 1779. Vol. II: Thomas Jefferson, June, 1779-June, 1781. Richmond, 1926-1928.

Virginia Historical Register. Richmond, Va., 1848-1851. 4 vols.

Virginia Magazine of History and Biography. Richmond, 1893-, vol. 1-

Wertenbaker, T. J. *Norfolk; Historic Southern Port*. Durham, N. C., 1931.

William and Mary College Quarterly. Williamsburg, 1892-1943. Series 1, vols. 1-27; series 2, vols. 1-23.

Williams, M. R. *Sailing Vessels of the Eighteenth Century*. (In *United States Naval Institute Proceedings*, vol. 62, January, 1936.)

Winsor, Justin. *Narrative and Critical History of America*. Boston, 1886. 8 vols.

Wise, J. C. *Ye Kingdom of Accomack, or The Eastern Shore of Virginia in the Seventeenth Century*. Richmond, Va., 1911.

Wissler, Clark. *Indians of the United States*. Garden City, 1946.

APPENDIX I

The following advertisements of vessels TO BE SOLD were selected from the *Virginia Gazette* as showing types and sizes of watercraft in use.

1739, MAY 4. . . . a small shallop about five years old in Yorktown, will carry between 400 and 500 bushels of corn.   William Rogers.

1745, . . . by the executors of Mr. Thomas Rawlings, a ship carpenter, lately deceased, the frame of a snow which was to have been built by the said Rawlings on account of Mr. John Hood, merchant, of Prince George County, of the following dimensions: 60 feet keel, 23 feet 8 in. beam, molded, 10 feet hold, 4 feet between decks. To be sold at the plantation of the deceased near Flower de Hundred. Also, a sizable, useful boat and a vessel called a schaw.

1745, JUNE 18. . . . To the highest bidder, schooner belonging to the estate of the Rev. Adam Duckie, deceased, trimmed and well-fitted with sails and rigging, some parts new, close docked, carries 50 hogsheads of tobacco . . . Also, a 12 hogsheads flat lying at Hobb's Hole.

1746, MARCH 27. The sloop *Little Betty* lying at Suffolk town in Nansemond county, burthen 50 tons, with her sails, anchors, furniture, tackle, will be sold on Wednesday, 9th of April.

1751, SEPTEMBER 26. . . . by the subscriber living in Norfolk county, a new schooner, now on the stocks and will be launched by the last day of November next, or sooner if required; the dimensions, 49 feet keel, 21 feet beam, 9 feet 6 inches hold. She is a well built vessel, her plank being well seasoned and sufficiently secured with iron work, being to be finished to a cleat, at 50 shillings per ton.   William Ashley.

1754, JUNE 20. . . . the brig *Lucy and John*, burthen 80 tons together with guns, rigging, tackle, apparel and furniture, at York Town, Friday, the 26th instant, to the highest bidder.   Thomas Dickinson.

1755, MAY —. . . . at public auction May 22, at the landing of Mr. Thomas Scott in the borough of Norfolk, a new ship on the stocks, dimensions: 62 feet keel, 23 feet beam, 11 feet hold, and 4 feet 6 inches 'tween decks.   Joshua Corprew.

1766, JUNE 27. . . . at Norfolk, a ship on the stocks, dimensions: 63 feet keel, 23 feet beam, 9 feet 8 inches hold, 4 feet 4 inches between decks, together with the rigging, sails, cables, anchors, etc., provided for her. She will be completely furnished and ready to launch by the 20th of next month. For terms apply to   Thomas McCullock.

1766, SEPTEMBER 19. . . . On the 16th day of October next at public auction to the highest bidder . . . a new ship about 170 tons burthen, well calculated for European or West Indies trade, and built with the best white oak complete and ready for launching with the full stock and rigging complete. Apply to administrators in Norfolk for William Irving.

1766, September 26. To be let on charter for Europe the snow *Nancy*, John Ardis master, now lying at Norfolk, a new vessel, burthen about 270 hogsheads. Apply to John Greenwood.

1766, November 6. . . . a new ship, 180 tons, built of white oak, for the West Indies or tobacco trade. Apply to Joseph Calvert, or to George Walker at Hampton.

1767, May 7. . . . a new ship now lying at Suffolk wharf, burthen about 350 hogsheads of tobacco, well built with best white oak timber and plank. The purchaser may have long credit for part of the money. Any person inclinable to purchase may be shown the vessel by applying to subscriber, living in Kingston Parish, Gloucester county. Thomas Smith.

1767, May 11. . . . a new ship of about 236 tons, well calculated for the tobacco trade, built of the best seasonal plank and timber, and can be launched in a little time, if desired. Two month's credit will be allowed for two-thirds or three-fourths the value. Any person inclinable to purchase may be shown the vessel by applying to subscriber, living in Kingston Parish, Gloucester county. Thomas Smith.

1768, March 15. . . . a well built snow, carpenter's and outside work finished, dimensions: 51 feet keel, 21 feet beam, 9 feet clear lower hold, 3 feet 6 inches between decks. Norfolk, executors of Joshua Nicholson.

1768, June 9. . . . a new schooner that will be launched in August next or sooner if required; burthen 71 tons, and will carry about 3000 bushels of grain; built of the best white oak plank and timber. Also, for sale, a sloop, 25 tons, one year old, together with her sails, anchors, etc. Apply to Edward Hughes, living on the head of East river in Gloucester county.

1768, June 16. . . . at Rocket's Landing, one-third, one-half or the whole of a schooner to be launched in a fortnight.   Samuel du Val.

1768, August 4. . . . a sea schooner, 80 tons, two years old. Also a sloop, 50 tons, now on the stocks, launched in three weeks. Kingston Parish, Gloucester county. Robert Billings.

1768, August 28. . . . a new vessel on the stocks, double decked, about 300 tons, might be launched in 24 days. John Greenwood, Norfolk.

1768, September 29. . . . a new vessel now on the stocks, of about 176 tons, tobacco or West Indies trade, built of the best seasoned plank, and can be launched in a few weeks. She may be made a ship, a snow, or a brig as may best suit the purchaser. Apply in Norfolk.

Edward H. Moseley

1768, OCTOBER 20. . . . . a double decked vessel on the stocks, 110 tons, will carry a great burden and is esteemed a very fine vessel.

Benjamin Harrison.

1770, MARCH 7. . . . the brig *Little Benjamin* about 110 tons burthen, double decked, has made but two voyages, is extremely well built and completely fitted. Credit will be given until the 10th of December next on giving bond with a good security to Ben: Harrison.

1770, MARCH 11. . . . anytime between this and the 10th of April next, the brigantine *Fair Virginian,* only one year old, just sheathed and now ready for to take a cargo on board, burthen about 100 tons. Any person inclinable to purchase such a vessel may know the terms by applying to the subscriber in Charles City and be shown the said vessel now lying near Sandy Point on James river. Cash or bills of exchange any time in the April General Court, will be accepted for payment.

Robert McKittrick, William Acrill.

1770, APRIL 13. . . . ready to launch being completely finished, a schooner, 41 feet keel, 18 feet 4 inches beam, and 8 feet hold; her beams, carlings, and top timber of cedar, and built by a compleat workman. Any person in want of such a vessel may be supplied by the subscriber on paying one-half the purchase money on delivery of said vessel, and the other half in October next. Also, a sloop, burthen of about 4000 bushels, will be ready by the first of May, and wants a freight for any part of the West Indies. Any person in want of such a vessel is desired to make it known to Carter Tarrant.

1776, SEPTEMBER — . . . the sloop *Industry,* now lying at Fredericksburg, with her sails, rigging, etc. She will carry upwards of 4000 bushels of grain. J. Watson and R. Dickinson are authorized to sell her.

Although the following contracts for building vessels were made when Virginia was no longer a colony but had become a state, they are included here because of the descriptions of the vessels and the interesting contracts:

(1) Contract between the owner and builder of a vessel in Gloucester county on July 31, 1777:

It is this day agreed on between Mathias James of the one part and John Fowler of the other part . . . That the said Mathias James for and in consideration of the sum of 35 pounds to him in hand paid, the receipt whereof he hereby acknowledgeth, doth oblige himself to begin, finish, and complete all the joiner's work properly belonging to the sloop he is

67

now building, in a neat, convenient and workmanlike manner. The steerage must be sealed that the whole shall be finished as soon as possible. In witness whereof we have hereunto set our hands and seals, the day and year above written. N. B.—There is to be no State Room in the above cabin.   Matthew James, John Fowler.
                                                        Witness, William Lilly.

(2) Contract between the owner and builder of a vessel on November 20, 1779:

I, Joseph Billups, Sr., of Gloucester county, Kingston Parish, do agree to build a boat 34 feet keel, with proper width of beam and hold, for John Avery . . . I do hereby oblige him first to pay me, the said Billups, 120 gallons of good West India rum, and 300 pounds of lawful money . . . The said Avery to oblige himself to pay the said Billups 100 pounds per ton, to supply the said Billups with suitable iron at ten shillings per pound . . . To furnish him with money if wanting to carry on the said boat . . .   Joseph Billings, John Avery.
                                                    Teste, Joseph Billups, Jr.

Various statistics were given by different writers for the number of Virginia owned vessels in the period just before the outbreak of the Revolutionary War. In *Shipyard Statistics* by H. C. Smith and L. C. Brown, one of the articles that comprises *The Shipbuilding Business in the United States of America,* edited by F. G. Fassett, Jr., and published in 1948 by the Society of Naval Architecture and Marine Engineers, there are given lists of vessels owned by the several provinces in the years 1769, 1770, and 1771. Virginia is listed as having in 1769, 6 ships, 21 sloops and schooners—27 vessels of 1269 tonnage; for 1770, there were 6 ships, 15 sloops and schooners, 1105 tons; and for 1771, 10 ships, 9 sloops and schooners, 1678 tons. We notice that the report of 27 vessels for 1769, is the same number reported by Governor Andros in 1698, which is rather surprising, and shows how inadequate the statistics were, and how careful a writer must be in using them.

APPENDIX II

The items on shipping given below were selected from the *Virginia Gazette* to show some details of Virginia shipping in the eighteenth century: the home ports, the ports entered and cleared, the types of vessels and various kinds of cargo. Sailings are given from September 3, 1736, when a Virginia owned vessel was first mentioned in the *Gazette,* to June 28, 1768, and is by no means a complete list, even in the copies of issues now extant; it is well to recall that copies of many issues have never been found. Later sailings in the *Gazette* have frequently omitted the type of vessel. A large number of vessels here named were Virginia owned and many of them Virginia built.

1736, SEPTEMBER 3. Ship *Priscilla* of Virginia, Richard Williams, entered at the port of York river from Barbadoes.

1736, NOVEMBER 9. Ship *John and Mary* of Virginia, Richard Tillidge, entered the port of York river from Barbadoes.

1737, FEBRUARY 9. The brigantine belonging to Col. Benjamin Harrison, arrived in James river last week from London, but last from Salt Islands loaded with salt.

1737, FEBRUARY 9. Cleared out of York river the schooner *Grampus,* John Briggs, for Madeira with 870 bu. wheat, 1451 bu. white pease, 1914 bu. red pease, 40 bu. beans, 1 hhd. beeswax, and 600 staves.

Cleared out of York district the following vessels:

1737, MARCH 2. Sloop *Medford* of New England, James Hathaway, for New England with 1000 bu. corn, 100 bu. pease, and 600 ft. of walnut plank.

MARCH 3. Ship *Hanover* of Bristol, Roger Rumney, for Bristol with 294 hhd. tobacco, 50 tons iron, and 5280 staves.

1737, MARCH 3. Schooner *Swallow* of New England, John Atwood, for Boston with 1500 bu. corn, 100 bu. pease, 20 bu. wheat, and 60 ft. of plank.

1737, MARCH 14. Sloop *Francis* of Bermuda, William Mallory, for Bermuda, with 2000 bu. corn, and 30 bu. pease.

1737, MARCH 18. Sloop *Mary* of Bermuda, Samuel Nelms, for Bermuda, with 5000 bu. corn, 56 bu. pease, 1 mast, and other pieces of timber.

1737, MARCH 19. Ship *Micajah and Philip* of London, James Bradley, for London, with 734 hhd. tobacco, 7500 staves, and a parcel of plank.

1737, MARCH 31. Brig *Abington* of Virginia, John Upcott, for Madeira, with 1170 bu. pease, 1617 bu. corn, 162 bu. wheat, beeswax and hemp.

Entered in the York District, with sundry European goods:

1737, MARCH 4. Ship *Catherine* of London, William Taylor, from London.

1737, MARCH 9. Ship *Haswell* of London, John Booch, from London.

1737, MARCH 18. Sloop *Southampton* of London, Robert Angus, from London.

1737, MARCH 23. Sloop *Betty* of Virginia, Thomas Hamlin, from Jamaica.

1737, APRIL 22. The ship *Johnston* of Liverpool, James Gillart, is lately arrived at York from Angola, with 490 choice young slaves. The sale of them began on Tuesday the 12th instant, and continues at York river. Thomas Nelson.

1737, MAY 2. Entered York river schooner *Lark* of Virginia, John Thompson, from Jamaica with 31 casks molasses, 6 puncheons rum, 3 bags cocoa, and 200 pounds [sterling] in cash.

1737, MAY 12. Entered York river, the sloop *Molly* of Virginia, Simon Handcock, from Barbadoes, with 32 hhd. 64 tierces and 70 bbl. rum, 61 bbl. sugar, and 1 bag ginger.

Cleared from Upper District of James river:

1737, JUNE 16. Sloop *Betty* of Virginia, George Cabanis, for Bermuda, with 764 bu. corn, 60 bbl. pork, 10 bbl. beef, 7 bbl. tallow, and 3 bbl. lard.

1737, JUNE 17. Sloop *Phoenix* of Virginia, Lemuel Portlock, for Barbadoes, with 696 bu. corn, 144 bu. pork, and 7000 staves.

1737, JUNE 18. Sloop *Molly* of Virginia, John Thompson, for Barbadoes, with 2534 bu. corn, 182 bu. pease, 38 bbl. pork, 1000 headings, and 4000 shingles.

1737, JULY 1. Entered York District, the brig *Priscilla* of Virginia, Richard Williams, from London and Madeira with 23 pipes and 1 hhd. Madeira wine.

1737, JULY 18. Entered York District the sloop *Industry* of Virginia, John White, from Maryland; cleared for Maryland with 400 bbl. salt and 7 doz. bottles Madeira wine.

1737, JULY 29. Cleared from York river the brig *Mary* of Virginia, Stephen Swaddle, for London with 105 hhd. tobacco, 1000 staves, a

70

parcel of sassafras, 13 pipes Madeira wine, 16 lbs. beaver skins and 6 doe skins.

1737, SEPTEMBER 17. Cleared out of York river, the brigantine *Priscilla* of Virginia, John Langland, for Bristol with 126 hhd. tobacco, 7 bbl. turpentine, 18 tons iron, 47 walnut planks, 49 gum planks, 7350 staves, and 1 bag wool.

1737, OCTOBER 28. Entered York river, the sloop *John and Mary* of Virginia, J. Briggs, from St. Christophers with 5 tierces, 1 hhd. molasses, 600 bu. salt, and 102 pounds [sterling] in cash.

1737, DECEMBER 9. The brigantine *John and Mary*, Richard Tillidge, now lies at Mr. Littlepage's wharf on Pamunkey river ready to take in tobacco on freight at the usual rate for Bristol. It is intended to sail in March. Orders sent to Captain John Perrin, owner, of Gloucester or Captain Tillidge.

1737, DECEMBER 16. The ship *Industry*, John Brown, now lying at Bull Hill in James river, will sail shortly for Cadiz, and is to call at Madeira in his return thither for wine and freight if sufficient encouragement is shown. Send orders to Captain John Hutchins of Norfolk, the owner of the ship, or to the master.

1738, MAY 1. Entered York river, the sloop *Molly* of Virginia, John Thompson, from Jamaica, having on board 45 casks molasses, 200 gal. rum, 1 hhd. sugar, 1 bag ginger, and 100 pounds in cash. She belongs to Captain Francis Willis.

1738, MAY 1. Entered York river, the sloop *Coan* of Virginia, John Kerr, from Dublin, having on board 1 chest linens, provisions, and 53 passengers. She is in the employ of Colonel Martin, who arrived in her.

1738, JUNE 7. Cleared from Upper James, the snow *Phoenix* of Virginia, William Spry, for London with 200 hhd. tobacco, 5 hhd. skins, 4 hhd. ipecacuane, 1 box sundry goods returned, 6000 staves, and 1 hhd. sassafras.

1738, JUNE 12. Entered York river, the brig *Abingdon* of Virginia, Thomas Southwick, from Barbadoes with 6 hhd., 80 tierces and 116 bbl. rum, 42 bbl. sugar, 16 hhd. and 1 tierce molasses, and 2 bbl. ginger.

1738, JUNE 30. The schooner *Fanny* lying at Mill creek near Hampton, will soon be higher up the James. Persons apply for freight to Mr. Jacob Walker or to Messrs. Cherrington and Whitten near the Falls of James river.

1738, JUNE 30. Goods on board the ship *Harrison* at Swinyards in James river, Thomas Bolling, owner of goods unknown. Any person sending for them with bills of lading may have them.

1738, JULY 27. Entered in York river the sloop *Molly* of Virginia,

John Thompson, from Barbadoes with 45 hhd., 8 tierces, and 9 bbl. rum, 69 bbl. sugar, 1 bag cotton, and 3 Negroes.

1738, July 28. A ship belonging to Mr. Theophilus Pugh of Nansemond is lately arrived in Nansemond, 13 weeks from Bristol.

1738, August 7. Entered Upper District of James river, the brigantine *Little Molly* of Virginia, Thomas Hamlin, from Jamaica with 7 hhd. sugar, 8 puncheons rum, 4 bags and 3 casks of cocoa.

1738, August 17. Cleared at York the schooner *Grampus* of Virginia, John Briggs, for Boston with 900 bu. pease, 600 bu. corn, 180 bu. wheat, 400 ft. walnut plank, 300 pipe staves, and 1 hhd. Madeira wine.

1738, October 4. Cleared from York the ship *Harrison*, Captain Bolling, for London.

1738, October 26. Arrived in York river the schooner *Grampus* of Virginia belonging to Colonel Lewis of Gloucester, John Briggs, from Boston with 6 bbl. cider, 5 bbl. train oil, 6 bbl. codfish and mackerel, 1 cwt. iron, 4 bbl. cranberries, 30 bu. apples, 1 tierce molasses, 5 hhd. and 6 bbl. rum, a Negro slave and 250 lb. cheese.

1738, October 26. The snow *Catherine and Lenora*, James McCullock, belonging to Messrs. Spaulding and Lidderdale, loaded with tobacco and bound for London, will sail from James river in 3 or 4 days.

1738, October 27. Arrived in York river last Monday the snow *John and Mary* belonging to Captain John Perrin, Richard Tillidge, from Bristol.

1738, October 28. Cleared from Upper District of James river, the sloop *Nancy* of Virginia, James Griffin, for Boston with 1307 bu. wheat, and 153 deer skins.

1738, November 6. Cleared from Upper District of James river, the snow *Kitty and Nora* of Virginia, James McCullock, for London with 223 hhd. tobacco, 16 casks skins, 1 parcel beaver skins, 4200 staves, and 400 ft. oak plank.

1738, November 13. Cleared out of Rappahannock District the ship *Brothers*, Robert Hall, for London with 471 hhd. tobacco, 40 tons pig iron, and 7000 staves.

1738, November 23. Cleared out of York District, the ship *Molly* of Virginia, Thomas Wilson, for Madeira with 1014 bu. wheat, 130 bu. corn, 107 bu. bonnevelts, 2 hhd. and 2 bbl. beeswax, 4 bbl. flour, and 100 hhd. staves.

1738, November 23. Cleared out of Upper District of James river, the sloop *Charming Anne* of Virginia, Thomas Goodman, for Lisbon with 3765 bu. wheat.

72

1738, DECEMBER 6. Entered in the Upper District of James river, the snow *John and Mary* of Virginia, Richard Tillidge, from York river in ballast.

1738, DECEMBER 9. Cleared from York river the schooner *Grampus* of Virginia, John Briggs, for Madeira with 2300 bu. of wheat, 1200 pipe staves and 143 lb. beeswax.

1739, JANUARY 1. Cleared from York river the brig *Abingdon* of Virginia, Thomas Southwick, for Madeira with 2709 bu. wheat, 152 bu. pease, 112 bu. corn, and 2000 lb. bread.

1739, JANUARY 26. Cleared out of Upper District of James river, the brig *Little Molly* of Virginia, Thomas Hamlin, for Georgia with 2551 bu. corn, 269 bu. pease, 33 casks pork, 8 casks beef, 2 casks lard, 8,314 shingles, I Negro, and 30 sheep.

1739, JANUARY 29. Entered the Upper District of James river, the brigantine *Robert and John* of Virginia, John Cooke, from the Lower District in ballast.

1739, JANUARY 30. Cleared out of Upper District the snow *John and Mary* of Virginia, Richard Tillidge, for York river with 4977 bu. wheat.

1739, FEBRUARY 4. Cleared out of York river the snow *John and Mary*, Richard Tillidge, bound for Madeira, having on board 4977 bu. wheat, 144 bu. pease, and 2000 lb. bread.

1739, FEBRUARY 5. Entered in the Upper District of James river, the sloop *Nancy* of Virginia, James Griffin, from Rhode Island with 6 bbl. train oil, 545 lb. cheese, 9 hhd., 8 tierce rum, 4 hhd., 4 tierce molasses, and a bundle of European goods.

1739, MARCH 8. Cleared out of James river, the brig *Robert and John* of Virginia, John Cooke, for Madeira with 5400 bu. wheat.

1739, MARCH 9. Cleared out of James river the sloop *Robert* of Virginia, Samuel Rogers, for Barbadoes, with 47 bbl. pork, 800 bu. corn, and 53 bu. pease.

1739, MARCH 23. Last Friday, the brig, *Pretty Betsy* belonging to Colonel Lewis of Gloucester county, James Robinson, bound for London with 202 hhd. tobacco, sailed out of Severn river and on the same day met with disaster on the Middle Ground between the Capes.

1739, MAY 3. Entered in York river the brig *Pretty Betsy*, Anthony *Mosely*, for London with 202 hhd. tobacco, 5000 staves, 1 pipe Madeira wine, and 22 tons iron.

1739, MAY 21. Entered Upper District James river, the snow *Kitty and Nora* of Virginia, James McCullock, from London via Madeira with sundry European goods and 12 pipes, 1 hhd. Madeira wine.

1739, MAY 21. Entered in York river, the brig *Abingdon* of Virginia,

Thomas Southwick, from Madeira and Barbadoes with 10 pipes wine, 15 hhd., 50 tierces and 63 bbl. rum, 37 bbl. sugar, and 9 pounds 8 shillings in cash.

1739, June 1. Cleared from York river the schooner *Grampus* of Virginia. John Briggs, for Madeira with 2460 bu. corn, 80 bu. pease, 1200 pipe staves, and 150 pounds beeswax.

1739, June 4. Entered the Upper District of James river, the ship *William and Betty* of Virginia, John Turner, from the Lower District with 323 hhd. tobacco.

1739, June 14. Entered in York river, the snow *John and Mary* of Virginia, Richard Tillidge, from Madeira and Barbadoes with 98 hhd., 21 tierces and 20 bbl. rum, 86 bbl. Muscavado sugar, and 12 pipes Madeira wine.

1739, June 16. Entered York river the snow *Mary* of Virginia, James Hume, from James river with 64 bbl. pork, 5600 shingles, 4200 pipe staves, and 4200 ft. 1-inch plank.

1739, June 22. The snow *John and Mary*, Richard Tillidge, belonging to Captain Perrin, now lying at Mr. Littlepage's on Pamunkey river, is ready to take on freight for Bristol.

1739, July 6. Cleared from Upper District the snow *Kitty and Nora* of Virginia, James McCullock, for London with 228 hhd. tobacco, 9 hhd. skins, 182 deer skins, 149 beaver skins, 56 walnut planks, and 4200 staves.

1739, August 11. Entered York river the brig *Little Molly* of Virginia, James Cox, from James river with part of her lading for the West Indies.

1739, September 8. Cleared York river, the brig *Abingdon* of Virginia, Thomas Southwick, for Madeira with 1861 bu. wheat, 1096 bu. corn, 118 pounds beeswax, and 1 case cloths.

1739, November 30. Last Saturday arrived in James river the sloop *Charming Anne* belonging to Colonel Benjamin Harrison, Captain Taylor, from Jamaica. Left James river for Jamaica on June 25, with 4000 staves, 487 bbl. pork, 37 bbl. beef, 2 bbl. tongue, 15 bbl. lard, 58 bbl. flour, 250 bbl. pease, and 70 bu. corn.

1745, April 12. Cleared at Hampton, the snow *John and Mary*, Thomas Bradley, for Liverpool with 106 hhd. tobacco, 500 bbl. tar, 60 walnut stocks, and 5600 staves.

1745, April 19. Entered at Hampton, the sloop *Little Molly*, Crawford Conner, from Philadelphia.

1745, May 17. Entered Hampton, May 3 to 17, 7 vessels.

1745, December 4. Cleared Upper District from September 20 to December 4, 14 vessels.

74

1745, December 27. Entered Upper District from September 20 to December 27, 20 vessels.

1746, July 31. Entered York river the snow *Two Brothers*, with upwards of 200 fine healthy slaves, the sale of which will begin at West Point on Monday, 4th of August. The said ship is not two years old, well-fitted and manned, and will take in tobacco for Bristol at 14 pounds per ton. Such gentlemen as are inclined to ship to Thos. Chamberlayne & Co., from York or James river, are requested to send their orders on board to John Lidderdale.

1746, July 31. Arrived from Gambia, the ship *Gildart* with 250 choice Gambia slaves, the sale whereof will begin at Hobb's Hole on the Rappahannock, on Tuesday, Wednesday, Thursday, the 5th, 6th, 7th of August; and in Brown's church the Monday following, where the sale will continue until completed. The said ship is a new vessel mounted with 20 guns, navigated with 45 men, and will take on tobacco for Liverpool at 14 pounds per ton. Apply to John Lidderdale, Harmer & King.

1751, January 1. Entered in York river the snow *London* of Virginia, Alex Leslie master.

1751, January 14. Cleared from York the sloop *Merry Fellows*, Thomas Perrin, for Barbadoes.

1751, January 18. Cleared from York the snow *London* of Virginia, Alex Leslie master.

1751, January 24. Cleared from York the snow *John and Mary*, of Virginia, Anthony Allen.

1752, September 21. Cleared from the Upper District of James river: (1) the ship *Bobby of Virginia*, John Cook, for London with 322 hhd. tobacco, 20 tons pig iron, and 7500 staves. (2) The snow *Phoenix* of Virginia, Samuel Kelly, for London, with 238 hhd. tobacco, 22 elephant's teeth, 1400 staves, 3200 heading, 50 pine planks, 100 hand spikes, and 14 oars.

1752, November 4. Cleared from the port of South Potomac, the *Caple* of Virginia, Samuel Curle, for Hampton, with 300 bu. Indian corn, 30 casks molasses, 17 bbl. and 6 tierce sugar, and 5 hhd. rum.

Entered at the port of Accomack the following vessels:

1768, May 13. Schooner *Anne*, William Wainhouse, from New York with 2 boxes chocolate, 800 wt. ham, 6 bbl. cordial, 3 cases and 2 half-bbl. rum, 6 cases and 1 bbl. loaf sugar, 1 quarter box glass, 6 hhd., 3 tierces, and 1 bbl. molasses.

1768, May 17. Sloop *Nancy*, Johannes Watson, from Philadelphia.

1768, May 18. Sloop *Endeavor*, Edmund Joyne, from Maryland.

1768, MAY 31. Schooner *Betsey and Esther*, Stephen Sampson, from Barbadoes with 24 hhd. rum, and 13 bbl. Muscavado sugar.

1768, JUNE 6. Sloop *Nancy*, Johannes Watson, from Philadelphia with 200 bu. salt, and a parcel of earthen ware.

1768, JUNE 10. Schooner *Little Betsy*, Zephaniah Brown, from Rhode Island, with one-half ton hollow iron ware, 2 hhd. rum, 20 bu. salt, a parcel of earthen ware, 2 riding chairs, 2 desks, 2 saddles, half-doz. house chairs, 2 trunks European goods, and 1 hhd. molasses.

1768, JUNE 11. Sloop *John and Betsey*, W. B. Hunting, from Philadelphia, with 1 box loaf sugar, 250 bu. salt, 2000 wt. cordage, 3 bbl. limes, 3 boxes European goods, 1 cask nails, 1 quarter-cask gun powder, 8 bolts duck, and a parcel of earthen ware.

1768, JUNE 13. Schooner *Jeany and Sally*, Reubin Joyne, from Nevis and St. Eustatia, with 7 hhd. rum, 1 hhd. molasses, 3 bbl. sugar, 3 hhd. foreign brown sugar.

1768, JUNE 20. Schooner *Old Plantation*, Laban Pettit, from Philadelphia, with 6 boxes chocolate, 2 boxes soap, 2 crates earthen ware, 4 saddles, 4 anchors, 3 doz. scythes, 1 bbl. loaf sugar, 2 tierces and 16 pieces of English duck, 1 trunk of European goods, 1 chest sweet oil, 1 cask nails, 3 kegs pipes, 1 tierce empty bottles, 1 box looking glasses, 2 bolts oznabrigs, and 1 piece sheeting.

Cleared at the port of Accomack:

1768, MAY 24. Sloop *Nancy*, Johannes Watson, for Philadelphia, with 1300 bu. corn, 5 bags feathers.

1768, MAY 28. Schooner *Friendship*, Daniel Sturgis, for Halifax with 3000 bu. corn.

1768, MAY 28. Sloop *Endeavour*, Edmund Joyne, for Boston, with 1600 bu. corn, and 200 bu. oats.

1768, MAY 28. Sloop *John and Betsy*, W. B. Bunting, for Philadelphia, with 1000 bu. corn, 20 bu. wheat, 60 bu. oats, 400 wt. feathers.

1768, JUNE 1. Schooner *Leah*, John Bradford, for Barbadoes, with 2000 bu. corn.

1768, JUNE 4. Sloop *Polly*, Thomas Alberton, for Philadelphia, with 900 bu. corn, 5 bbl. pork.

1768, JUNE 9. Sloop *Nancy*, Johannes Watson, for Philadelphia, with 1350 bu. corn, and 20 bu. oats.

1768, JUNE 9. Schooner *Skipton*, William Patton, for Maryland, with 700 bu. corn, 1000 wt. bacon, 2 cwt. feathers, 10,000 shingles.

1768, JUNE 27. Schooner *Old Plantation*, Laban Pettit, for Philadelphia, with 1200 bu. oats.

76

1768, June 28. Schooner *Little Betsey*, Zephaniah Brown, for Rhode Island, with 1650 bu. corn, 12 bu. wheat, 10 bu. pease, 10 bu. rye, 4 bags feathers, and 1 bag cotton.

An analysis of these items shows that the vessels entered and cleared at the York river, Lower James river, Hampton, Upper District of James river, Rappahannock, Pamunky, Nansemond, and Severn river. At least half of the entries and clearances were made in the York river. It will be noted that the same vessel made a number of entries and clearances. In the list are brigs, brigantines, sloops, schooners, snows, and ships, most of them Virginia owned, and we like to think they were Virginia built as well. Only six ships are listed as Virginia owned, yet the names of some of the others are so strictly Virginia names—*Braxton, Harrison, Virginia Planter*—that is seems highly probable that they too were Virginia owned. The names of only ten owners are given.

The information received by the *Gazette* was not always accurate. Occasionally a vessel is listed as two vessels of different rigs, but having the same name and the same master was evidence enough that they were one and the same. The *John and Mary*, Richard Tillidge master, is listed as a brigantine for two trips, a snow for eight trips, and a sloop, John Briggs master, for one entry. The *Robert and John*, John Cooke master, is listed both as a brig and a brigantine. Sometimes the name of a vessel was changed after its first appearance as in the case of the *Katherine and Lenora* which appeared on three trips thereafter as the *Kitty and Nora*, James McCullock master.

The cargoes of vessels clearing for Europe and the West Indies contained for the most part tobacco, corn, wheat, beans, pease, beeswax and staves. The cargoes from vessels entering from Europe would contain goods of various kinds; vessels from the West Indies would bring rum, molasses, sugar, ginger, salt, and occasionally a slave. In 1746, two ship loads of slaves were brought to the colony and sold, a part of the sale being conducted in a church.